SEVEN UNDERWATER WONDERS
OF THE WORLD

Trumpetfish, crimson soldierfish, and blue-striped snappers, Galápagos

Manta ray, German Channel, Palau

Dedicated to Susan Sammon

Published by Thomasson-Grant, Inc.
Designed by Leonard G. Phillips
Edited by Owen Andrews

99 98 97 96 95 94 93 92   5 4 3 2 1

Library of Congress Cataloging-in-Publication Data
Sammon, Rick.
Seven underwater wonders of the world / Rick Sammon.
p.   cm.
Includes bibliographical references (p.   ) and index.
ISBN 0-934738-78-5
1. Marine biology—Popular works. 2. Aquatic biology—Popular
works. 3. Marine resources conservation—Popular works. 4. Marine
biology—Pictorial works. 5. Aquatic biology—Pictorial works.   I. Title.
QH91.15.S36 1992
574.92--dc20   92-6413 CIP

Thomasson-Grant, Inc. • One Morton Drive • Charlottesville, Virginia 22901 • (804) 977-1780

SEVEN UNDERWATER WONDERS

OF THE WORLD

RICK SAMMON

THOMASSON-GRANT

Charlottesville, Virginia

Coral overhang, Red Sea

TABLE OF CONTENTS

INTRODUCTION

IN 1980, MY WIFE, SUSAN, AND I TOOK UP SCUBA DIVING. I HAD ALWAYS been fascinated by underwater exploration, but oddly enough, it was a job offer that finally pushed me into becoming a serious diver, when I was hired to produce a newsletter for CEDAM International, a marine exploration organization dedicated to Conservation, Education, Diving, Archeology, and Museums. To my delight, my new job required me to provide underwater photographs and firsthand accounts of CEDAM expeditions; together with Susan, I set out to learn the fundamentals of scuba journalism.

Over the next eight years, CEDAM International's marine biology and archeology expeditions took us to breathtaking underwater environments around the world, from the Florida Keys, Grand Cayman Island, and Belize Barrier Reef to the Red Sea, the Seychelles, the Kenya coast, and the Galápagos Islands. Before long, we had become directors of CEDAM International, organizing and leading the expeditions and running the office back home.

Our adventures with CEDAM gave us an intense appreciation for these wonderful marine environments. The fishes' remarkable colors and strange shapes fascinated us, as did their camouflage techniques, hunting methods, and other survival tactics. The times between dives were equally memorable. On the dive boats and on land, we met many people—guides, boat operators, fishermen—who taught us about reefs, weather, customs, and local environments, supplying insights that are simply not found in travel guides.

Everywhere we went, we encountered disturbing examples of human damage to this very fragile marine world. We soon knew we wanted to do something about it, but what that would be wasn't clear.

In 1988, after an unusually spectacular dive off the Kenya coast, I said to

The reefs of Wantamu Marine Reserve, Kenya, were among twenty-six exceptional sites discussed at the "Seven Underwater Wonders of the World" meeting.

Susan, "This site is simply wonderful." On the beach later that night under brilliant tropical stars, we discussed the idea of a conservation-oriented project that could spread the word about all the wonders of the underwater world and the need to protect them. We talked about a book, a television documentary, and a major traveling exhibition of underwater photographs. The word "wonderful" kept coming up as we fleshed out the project. It was dawn before we finished talking.

Upon returning home, we decided to launch our conservation idea. We called it the Seven Underwater Wonders of the World—realizing, of course, that the underwater world has far more than seven wonders. But we also knew that to get our message across, we needed a simple, appealing theme.

Making our list of wonders, we were tapping into a tradition over two thousand years old. In 130 B.C., a Greek poet named Antipater of Sidon composed a roster of seven wonders made by human hands: the Hanging Gardens and walls of Babylon, the Colossus of Rhodes, the statue of Zeus at Olympia, the Mausoleum at Halicarnassus, the Temple of Artemis at Ephesus, the Pharos Lighthouse at Alexandria, and the pyramids of Giza. Today, only the pyramids survive.

These ancient monuments are still the Seven Wonders to many people. But other lists have been created over the ages. One that appeals to me—the first to list natural rather than manmade wonders—was devised in the sixth century A.D. by St. Gregory of Tours, a Frankish bishop and historian. Gregory declared that Antipater's list was too pagan and offered his own list of six "Wonders Made by the Hand of God": the tides of the oceans, the growth of plants from seeds, the volcano Mount Etna, the rebirth of the phoenix, the yearly cycle of the sun, and the monthly cycle of the moon.

Like Antipater, twentieth-century listmakers have featured technological wonders. A list of "Modern Day Wonders" published by *Scientific American* magazine in 1913 cited airplanes, automobiles, reinforced concrete, the x-ray machine, phonographs, and motion pictures. *Current History* magazine's 1937 list included the New York City subway system; others mentioned the Empire State Building, the Eiffel Tower—and often, those enduring pyramids at Giza.

I learned about the Seven Wonders idea as a five-year-old in 1955 while doing my geography homework. That same year, Lowell Thomas, the noted

Ambergris Cay, Belize

explorer and adventurer, brought out the wide-screen movie *Seven Wonders of the World*, a film that thrilled me and millions of other moviegoers. While Thomas made sure to visit impressive structures like the Taj Mahal in India, he also seemed to share St. Gregory's ambition to make people appreciate natural wonders. His images of Africa's Victoria Falls and North America's Grand Canyon were my introduction to nature's magnificence.

The wonders theme intrigued me. It had staying power and the mystique of the number seven. Little did I know that someday it would dominate my life for several years.

By the summer of 1989, CEDAM International was ready to select the official Seven Underwater Wonders of the World. Our goals were simple. We wanted to show as many people as we could the beauty and fragility of the underwater environment. We wanted to initiate or enhance worldwide efforts to protect that environment. Recalling the disappearance of six of the world's original seven Wonders through damage or neglect, we wanted future generations to enjoy the beauty that lies beneath the ocean's surface.

"Seven Underwater Wonders of the World" selection committee. (Photograph © Susan Sammon)

To gain support for our idea, I worked on assembling a distinguished selection committee composed of people known for their marine research and environmental efforts. After numerous phone calls and letters, I was able, on August 25, 1989, to gather the following conservationists, scientists, and explorers in one room for one day: Mr. Jean-Claude Faby, United Nations Environmental Programme; Mr. Scott Carpenter, marine/logistics advisor and former NASA astronaut and aquanaut; Dr. Charles Carr, New York Zoological Society; Dr. Jacque Carter, Wildlife Conservation International; Dr. Eugenie Clark, University of Maryland; Mr. James Fowler, Fowler Center for Conservation; Lt. James Morris, National Oceanic and Atmospheric Administration; Dr. Robert E. Johannes, Center for Scientific and Industrial Research Organization, Australia; Mr. Emory Kristof, National Geographic Society; Dr. Ernest Ernst, New York Aquarium; Mr. Ian Koblick, Marine Resources Development Foundation; Dr. Andrew Rechnitzer, Viking Oceanographic; Ms. Marsha Sitnik, Smithsonian Institution; and Dr. William Stone, Cis-Lunar Development Laboratories. Actor Lloyd Bridges, who is dedicated to marine conservation and who starred in the 1960s television

The reefs of Indonesia are among the most spectacular in the world. Even though they were not chosen by the Seven Wonders selection committee, they, too, need to be protected and conserved.

program "Sea Hunt," was also on hand to lend his support to our project.

Each panelist made a presentation on the sites they felt were worthy of the title "Underwater Wonder of the World." Twenty-six sites were nominated, so choosing only seven was difficult. We judged each site on the following criteria: natural beauty, uniqueness of marine life, scientific research value, environmental significance, and geological significance.

After all the nominations were made and we had debated the merits of each, we designated the following places as our Seven Underwater Wonders of the World: the Belize Barrier Reef, Belize; Lake Baikal, Russia; the Northern Red Sea, Egypt; the Galápagos Archipelago, Ecuador; the Great Barrier Reef, Australia; the Deep Ocean Vents; and Palau, Micronesia.

Shortly after the meeting, letters of support started to come in from world leaders, including President Corazon Aquino of the Philippines, Prime Minister Bob Hawke of Australia, and President Oscar Sanchez of Costa Rica. Television and movie stars, including Ted Dansen, Jeff Bridges, and Candice Bergen applauded our efforts too.

We were thrilled at the media's warm response; more than a hundred articles appeared on the project—which was only an idea at that point. We had yet to launch its first expedition. These articles alone helped bring our conservation message to millions worldwide.

Seven months after the meeting, we embarked on the adventure of documenting the sites. In the following pages, we will take you to each, telling the story of the expeditions and surveying the present condition and future prospects for each wonder's survival.

Susan and I visited six of the sites together; I took the photographs, then wrote the first draft of the narrative between dives, while my impressions were as fresh as possible. The seventh site, the deep ocean vents, can only be reached in a special submarine named *Alvin*, at a cost of more than twenty thousand dollars a day. For the photographs and narratives from the vents, I turned to my friends Emory Kristof and Dr. Dan Fornari.

Marco Polo, one of the most famous explorers of all time, had an interesting theory about adventures. He wrote, "An adventure is misery and discomfort, relived in the safety of reminiscence."

After all my adventures with CEDAM International and the Seven Underwater Wonders of the World project, I know what the Venetian explorer meant. When I'm writing at home or speaking in an auditorium, I often make light of the difficulties I encountered in some far-off part of the world, where the weather was not always perfect, the seas were sometimes very rough, and underwater visibility often limited what we could do. These problems seem small in light of all that we saw and felt and learned. Why dwell on negative moments when the positive ones completely outweigh them?

As a conservationist, however, I've had to rethink that approach. At too many sites, the evidence of human damage was too widespread to be overlooked. I wondered how much the book should emphasize the damage that has already defaced many of our fragile underwater ecosystems. I finally decided that I could convey the beauty of the underwater world—and the need to protect it—through beautiful pictures of places that are still pristine.

There's one exception: the two images on this page, taken at the same time of day at the same spot in a remote part of the Indo-Pacific. After I took the colorful

Top *Healthy reef*

Above *Reef scene after dynamite fishing and live coral collecting*

picture at the top, I made a 180-degree turn and took the drab picture under it. Why the vast difference in color and life? The answer is twofold. One cause is dynamite fishing, where fishermen toss dynamite or homemade bombs in the water to stun the fishes and bring them to the surface by the hundreds, casually destroying sections of coral reef that take generations to grow back. The other is excessive collecting of corals and shells for the wholesale trade.

This scene is becoming all too common in the world's oceans. But the future holds promise. Conservation organizations are now developing education programs to help preserve the marine environment for future generations. Scuba divers and boat operators are becoming more aware of how important it is to stay off the reef. And marine parks are being created to protect and preserve these and other underwater wonders.

No photograph, however beautiful, or text, however lyrical, can convey the breathtaking quality of these environments—what life is like in the quiet, rhythmic surge of the waters that make up so much of our planet, and what makes each of the wonders wonderful.

I hope readers of this book will have the opportunity to visit many of the sites. I also hope that our project will contribute to the preservation of a unique part of our world.

Safe Diving.

Rick Sammon
Croton-on-Hudson, New York
1991

Scrawled filefish, Belize Barrier Reef

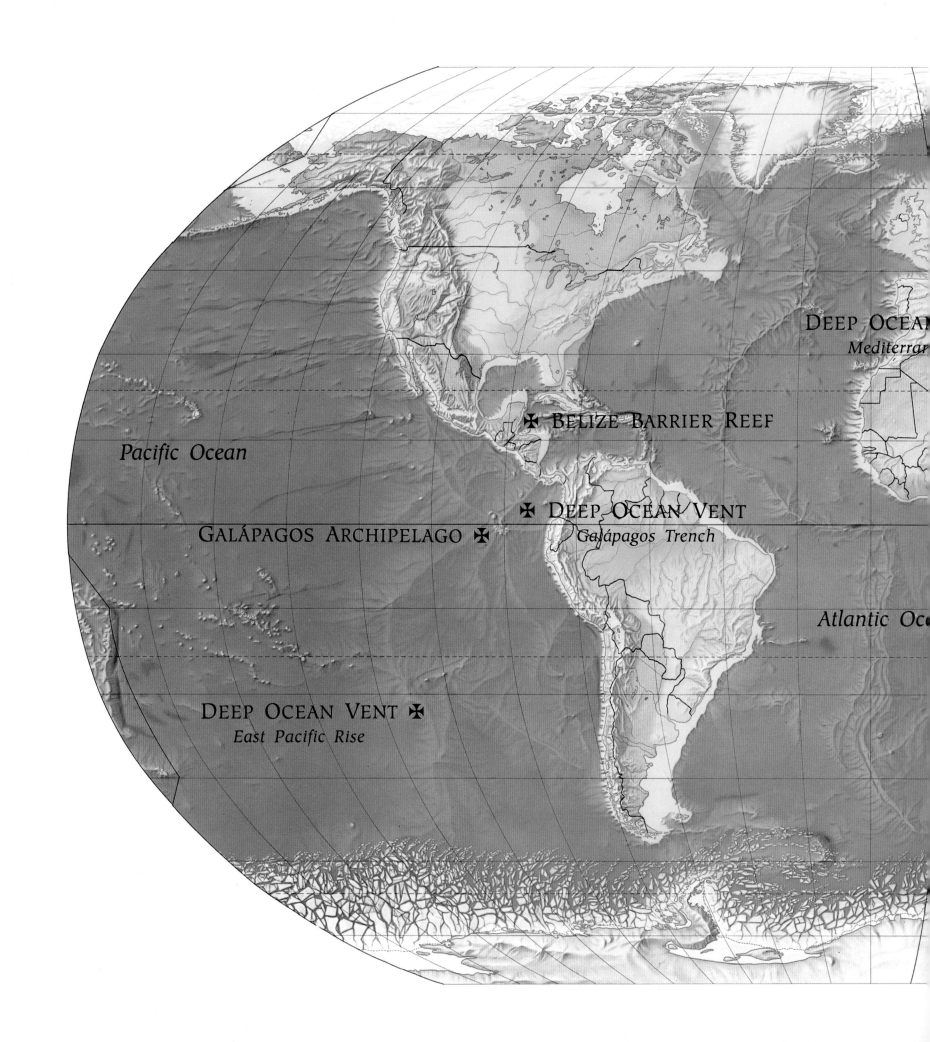

DEEP OCEAN
Mediterra

✠ BELIZE BARRIER REEF

Pacific Ocean

✠ DEEP OCEAN VENT
Galápagos Trench

GALÁPAGOS ARCHIPELAGO ✠

Atlantic Oce

DEEP OCEAN VENT ✠
East Pacific Rise

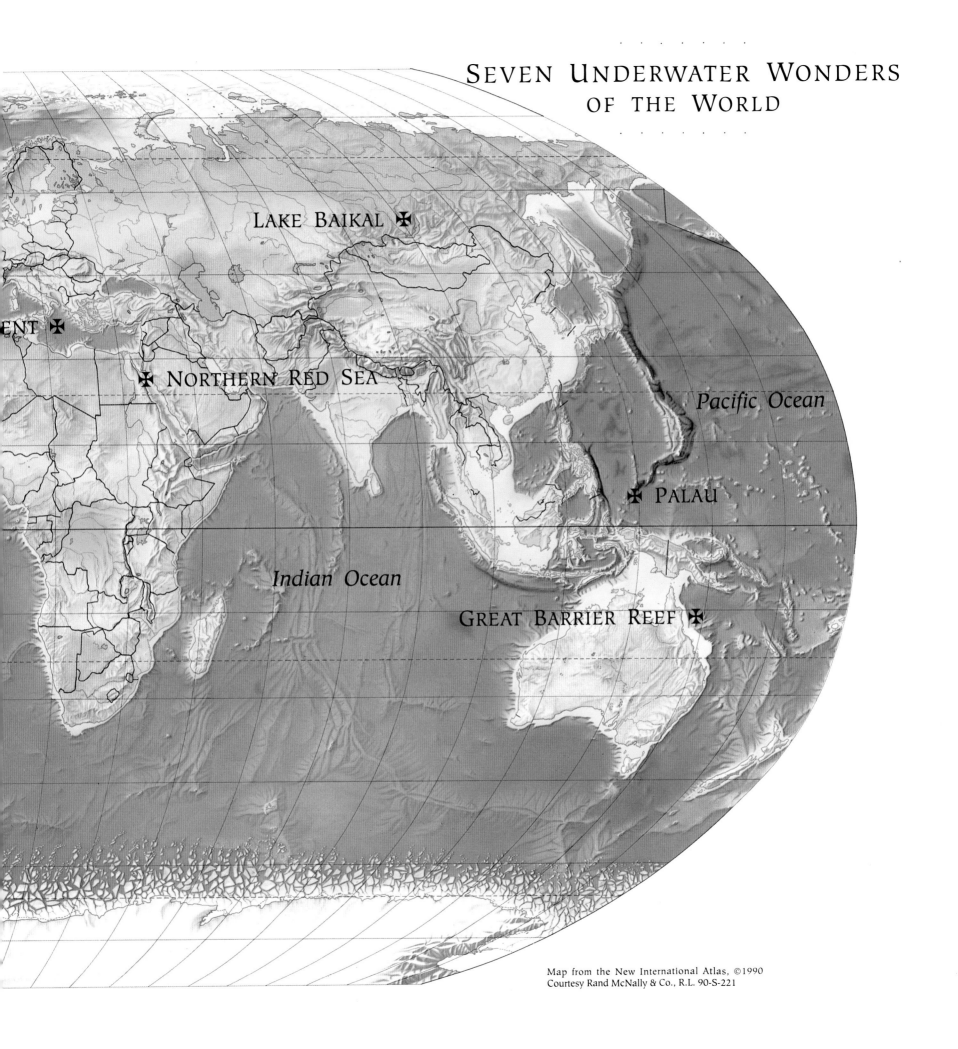

SEVEN UNDERWATER WONDERS
OF THE WORLD

LAKE BAIKAL ✠

ENT ✠

✠ NORTHERN RED SEA

Pacific Ocean

✠ PALAU

Indian Ocean

GREAT BARRIER REEF ✠

Map from the New International Atlas, ©1990
Courtesy Rand McNally & Co., R.L. 90-S-221

Belize Barrier Reef

WHAT IS THE REST OF THE WORLD DOING TODAY? I WONDER. IT'S 8:00 A.M., and my dive buddy and I are already sixty feet underwater, documenting the first site in CEDAM International's Seven Underwater Wonders of the World project.

We're beginning with the New World's largest known underwater wonder and the one closest to our base in New York: the Belize Barrier Reef, a 160-mile-long system of reefs along the coast of the small, English-speaking, Central American nation of Belize. Second only in size to Australia's Great Barrier Reef, this is the most diverse and luxuriant coral reef in the Western Hemisphere, stretching from the tip of Mexico's Yucatan Peninsula south into the Gulf of Honduras. As I write, the reef remains relatively intact and unspoiled, and as we travel from one dive site to the next, I will be finding out more about what is being done now to protect it.

I will be diving and photographing at Glover's Reef and Lighthouse Reef, two atolls east of the main reef. Atolls are rare in the Caribbean—there are only ten in the whole region—and three of these are in Belize. Most of the world's atolls are in the Pacific Ocean; there they form around volcanic islands, beginning as fringing reefs and gradually becoming rings around lagoons as the volcanic landmass sinks under the sea. Belize's three atolls have a somewhat different history, having anchored on the crests of offshore fault blocks.

Fifteen miles out to sea, Glover's Reef is an oval about a dozen miles long and five miles wide. Today we are at the southeastern cays ("cay" is the name Spanish explorers gave to the Caribbean's small, low islands), the most spectacular area for diving on the atoll. Here one finds both shallow and deep walls of colorful and thriving corals, home for a myriad of reef fishes.

The cays and reefs of Lighthouse Reef, an atoll forty miles east of the Belize coast, are plainly visible in this photograph taken from a space shuttle. Thirty miles long and five to ten miles wide, Lighthouse is a naturalist's delight; its unspoiled cays are home to tropical birds and trees, and its healthy reefs support a wide array of typical Caribbean marine species. (Photograph courtesy of NASA)

Facing *Divers will find spotted cleaner shrimps, which grow to a length of about one inch, on anemones throughout the Belize Barrier Reef. Protected from predators by the anemone's stinging tentacles, the shrimps keep their hosts free of parasites and silt.*

During the day, most hard corals keep their polyps withdrawn inside their cementlike external skeleton. At night, when many coral-feeding fishes are asleep, hard-coral colonies come alive, extending their tentacles to feed on zooplankton and phytoplankton.

As I explore the reef's outer face, I am swimming with fishes that look like their nonscientific names: butterflyfishes, hogfishes, and parrotfishes. Watching them, I am amazed at how they have adapted over time to survive in their environment. The foureye butterflyfish, for example, has a large, dark spot near its tail on each side of its body. To a predator, this spot may look more like an eye than the fish's actual eye and may cause it to mistake which way the butterflyfish will dart when threatened—possibly a life-saving chance for the butterflyfish.

Some species of butterflyfish can also change their coloration, as do many other reef fishes. At night, when they are sleeping in a reef corner or on the sandy bottom, they lose some of their brilliant daytime markings, becoming much less visible to predators.

With almost no effort, I swim over the dense coral garden that covers the sea floor, a convoluted backdrop of elkhorn, brain, boulder, and star corals. Tree-like soft corals, the gorgonians, sway gently in the light surge, spreading their branches to catch plankton. It is not easy to believe, as I look at this apparently peaceful mosaic, that hard corals compete vigorously with each other for light and space along the firm substrate or foundation of the reef, exuding poisons that kill their neighbors or simply growing fast enough to put nearby corals in the shade. The struggle is silent, gradual, and relentless.

It is easy to believe that the eighteenth-century naturalists who first attempted to classify corals thought they might be plants rather than colonies composed of thousands of living animals. Those naturalists were not entirely wrong, for hard corals are unique creatures. Within each coral animal live thousands of single-cell algal plants called zooxanthellae (pronounced zoo-zanthélly).

The relationship between coral animals and their algae tenants is symbiotic—essential to the survival of both parties—and extremely complex. This partnership is the foundation of the immense and intricate balancing act that is a living coral reef, a balancing act in which many hundreds of species participate.

In the clear, warm waters where coral reefs develop, nutrients are scarce. That is why the water is so clear and so blue; it isn't clouded with living plankton and algae. Like a forest growing and becoming mature over the centuries on a

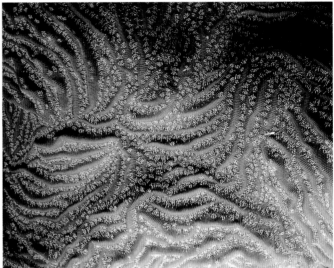

Like other species of hard corals, leaf corals secrete a mucus that removes sediment from their surfaces and prevents the polyps from being smothered. The rate of secretion is sufficient to remove moderate naturally occurring dustings. However, hard corals cannot handle the tremendous amount of sediment raised by human activities such as the dredging of the mangroves and seagrass beds at the southern end of the Belize Barrier Reef.

Left To help preserve the pristine reefs of Belize for future generations, visitors should follow the conservation-diver's creed: take only pictures and leave only bubbles.

Horse-eye jacks are swift swimmers and aggressive hunters. Often seen in schools of a dozen or more, they stalk schools of blue chromis. Jacks often carry ciguatera, a form of fish poisoning; travelers should avoid eating them.

rocky slope, the coral reef gradually builds up a rich ecosystem in a setting that would otherwise be quite sparsely inhabited.

This is possible in part because coral animals and zooxanthellae use the scant naturally occurring resources with remarkable efficiency. By day, the zooxanthellae, living along the walls of the coral colony, conduct their plantlike business—photosynthesizing sunlight, processing carbon dioxide and other wastes from the coral animals, and producing oxygen, sugars, and protein which the coral animals consume. At night, when the corals extend their polyps from openings in their hard calcareous skeletons, they trap and eat small planktonic animals, which will supply another round of waste for the zooxanthellae.

As the corals and zooxanthellae carry on with their exchange of nutrients, each coral polyp secretes calcium carbonate, or marine limestone, the stony structure of the colony. From time to time, new polyps develop from buds around the mouths of the old polyps. Slowly, through secretion and budding, the reef grows. Each hard coral species (there are several hundred worldwide) grows in a unique pattern, adding a particular shape to the body of the reef. Like the three-dimensional framework of a forest, the resulting structure provides countless places for other sea creatures to establish themselves.

I turn and glide over a nearly vertical wall, a drop-off to 110 feet. Looking into the deeper water, I see nothing but a foglike dark blue wall. I hear only the bubbles from my regulator. Between breaths, I can feel my heart beat. The rest of the world is indeed far away.

Occasionally an ocean trigger, a disk-shaped fish with permanently puckered lips, swims by. This is a relatively rare sighting, for ocean triggers are usually found in deeper offshore waters. Then, gently coming out of the dense blue fog, I see an Atlantic manta ray with a six-foot wingspan. He moves in slow motion, mouth open, straining plankton from the water as he flaps his massive wings. Gradually, this majestic animal disappears into the distance. I want to photograph him, but I learned long ago that it is impossible to outswim virtually any fish in the sea, large or small.

After ten minutes at eighty feet, I slowly ascend toward the shallows. Like all divers, I must be careful not to rise too quickly. Doing so would cause the compressed air in my body cavities to expand too rapidly, resulting in a painful or perhaps fatal embolism.

On the reef crest, where there is a greater variety of food, I find more fishes, especially juveniles. Groupers, grunts, snappers, filefish, and jacks are here, constantly on the lookout for food and alert for predators. Early-morning light is still streaking through the water at a forty-five-degree angle, silhouetting the curious fishes which are investigating this strange visitor to their realm. I enjoy diving in the early-morning hours, when nocturnal animals are returning to the safety of the reef's nooks and crannies and the diurnal fishes emerge. Like one watch on a crowded submarine, rolling out of their bunks to go on duty, they turn over their sleeping places to the nocturnal fishes.

Working quickly, as one must underwater, I begin to photograph the flora and fauna of this spectacular marine environment, hoping that luck, an essential element in underwater photography, is on my side, and that I will get what even the best underwater photographers admit to: perhaps one great shot per roll.

We have been underwater for almost an hour, at a remote site few divers get to see. Thanks to our wet suits, we are still comfortable in the 84° F water. My buddy taps on his tank to get my attention, and signals that we must return to our dive boat, the *Belize Aggressor.*

Although flaming scallops are bright red, they look brown at a depth of sixty feet, because color is filtered out selectively as depth increases. With an underwater strobe light, photographers can record the mollusk's true color on film.

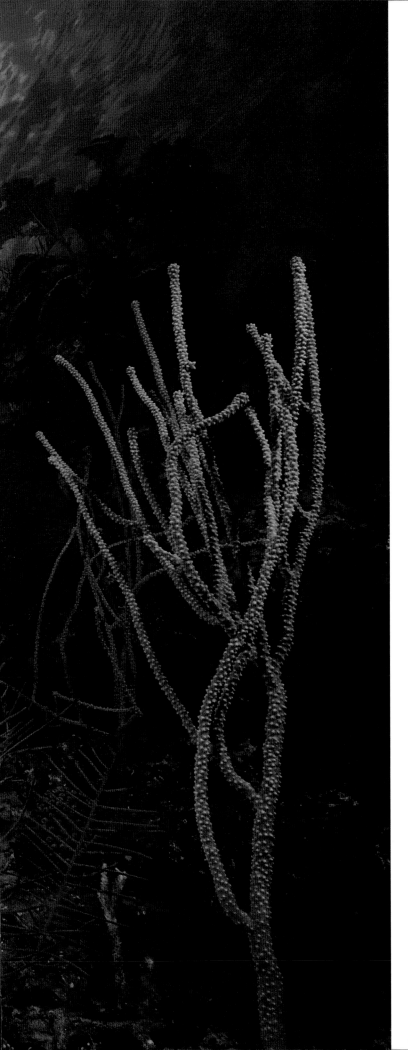

The scrawled filefish, a solitary swimmer, eats algae,
gorgonians, seagrass, tunicates, anemones, and corals.
A long spine on its forehead protects it from predators,
inflicting a painful puncture.

Left Hanging motionless amid soft corals, trumpetfish
wait for prey. By aligning their bodies and matching
their color with the coral branches, they become almost
invisible—revealed only by a diver's light.

The underwater ridges and valleys of barrier reefs protect coastal habitats by dispersing the energy of ocean waves. Exploring the reef's enormous variety of organisms, researchers hope to discover new sources of food and new medicines for humankind.

AT GLOVER'S REEF, I FIND IT EASY TO FORGET ABOUT TWENTIETH-CENtury civilization and its never-ending demands on the earth's resources. I think of fishermen in pre-Columbian times venturing out from small villages among the cays and lagoons of the barrier reef to try their luck near the atoll.

Those fishermen lived on the fringe of the Yucatan Peninsula's great Mayan civilization, which flourished from the third to the tenth centuries A.D. and had its origins in far earlier times. In the jungles of northern Belize, archeologists have found settlements at Cuello dating back to 2600 B.C.—some of the oldest evidence of human habitation in Mesoamerica. During the Mayan epoch, temple centers and towns appeared in Belize at Caracol, Las Milpas, Altun Ha, and dozens of other sites.

Trading and fishing villages emerged on Belize's cays and lagoons, and archeological evidence suggests a lively traffic between the inland towns, these outposts, and other parts of Yucatan; the Mayans cut a canal between Ambergris Cay and the mainland to avoid going around the reefs.

Even today the Mayan heritage persists in parts of Belize, evident in languages and customs that have survived almost five centuries of foreign influence. But the jungle long ago reclaimed the old cities, many of which remain unexplored (except, all too often, by looters), hidden in remote and roadless corners of the country. Belize is lucky that its rain forest remains relatively intact, still home to jaguars, five hundred species of tropical birds, and a multitude of reptiles including iguanas and the feared and deadly fer-de-lance snake.

Along Belize's shore grow dense mangrove thickets, tangled masses of tropical salt-water trees. Their existence here is helped by the barrier reef a few miles offshore, and the reef in turn benefits from the mangroves. Truly a barrier, the reef diminishes the force of ocean waves too strong for mangroves to withstand. For their part, the mangroves, with their labyrinthine root systems, lessen coastal erosion, especially during severe tropical rain storms, acting as a filter that traps sediment before it reaches and smothers the reef's living coral polyps.

The branches of the mangroves are alive with birds, reptiles, and insects. Their underwater roots are habitat for manatees, porpoises, loggerhead turtles,

The future of the Belize Barrier Reef worries Belize fisherman Villamar Godfrey. He feels the southern end of the reef is being hurt by chemical run-off from citrus farms and the dredging of the mangroves for deep-water harbors.

Dense mangrove thickets cover the southern end of
Ambergris Cay. Mangroves are essential for the reef's
survival. Their thick root systems trap silt and sediment before it reaches the reef and smothers corals. The
reef, in turn, protects the mangroves from strong wave
action.

tarpons, sponges, and many other marine creatures. The mangrove swamps are essential too for many reef fishes, which swim here to find breeding and feeding areas.

On a previous visit to Belize, I dove in the mangroves. Although visibility was only about four feet, I saw dozens of juvenile reef fishes, including sergeant majors and barracudas, dart between the sponge- and algae-covered mangrove roots. Exploring the mangrove channels, I found a few upside-down jellyfish, pulsing to nature's never-ending beat.

Between the mangroves and the barrier reef, the lagoons extend, five to ten miles wide, dotted with seagrass beds and patch reefs. Many reef fishes come here to graze during the day, while others cruise the area after the sun goes down. Here too, diving offers some interesting surprises—dozens of lobsters peering out of crevices, turtles dining on turtle grass, a tiny octopus scurrying across the sandy bottom.

The interrelated habitats of the mangrove swamps, the lagoons, and the reefs are the foundation for a wonderful abundance of fishes. For Belizeans, who mostly live along the coast, fishing has long been a way to survive. Their villages seem tranquil, simple, and picturesque to a New Yorker like me, and it worries me to hear that industrial and agricultural development on the coast and in the rain forest, overfishing of the reef, and even tourism could—all too soon—compromise the outstanding quality of Belize's coastal waters and threaten an old way of life.

With its rain forest and its reefs, Belize has much to protect. Both ecosystems seem so robust in their pristine state—deceptively so, because both are raised upon such slight foundations, the rain forest on thin soils that quickly wash away when the trees are cut, and the reef on clear and hence nutrient-poor water. When part of either habitat is destroyed, the loss will not be replaced in our lifetimes.

Glover's Reef is an isolated place, and conservationists think its chances of remaining unspoiled are good. Even here, though, petroleum companies have drilled exploratory wells, lured by oil discoveries off the coasts of Mexico and Guatemala. Currently, no laws protect any part of Glover's Reef, although it has

An underwater forest of mangrove roots provides unlimited places for hydroids, sponges, and algae to take hold. In this environment, where there are many "fast-food" opportunities, dozens of species of juvenile reef fishes are found, including sergeant majors and barracudas. Topside, the branches are crawling with crabs and filled with stinging insects.

Often seen riding the bow waves of dive boats on the Belize Barrier Reef, Atlantic bottle-nosed dolphins travel and hunt in groups, or "pods," numbering as many as a hundred individuals. Dolphins have highly developed cranial regions which may help them send and receive acoustic signals directed toward potential prey. Worldwide, there are thirty-two species of dolphins in seventeen genera.

Right *The brittle starfish is aptly named. When attacked, its arms easily break off. They begin to regenerate almost immediately. Brittle starfish are also known as serpent stars because of the sinuous way they move across the reef. Twenty-one species have been documented in Belize. Many live on, in, and around sponges such as this fire sponge.*

been proposed as a national reserve where fishing and tourism can be carefully regulated. Maintaining a reserve costs money and requires a lot of education so that the populace will support it—heavy burdens for the government of a small nation. As I relax between dives in the brilliant, cheerful weather, these thoughts remind me of the seriousness of my work with CEDAM International, and I find myself hoping that we can help make a difference.

.   .   .   .   .   .

LATER IN THE DAY, WE ARE CUTTING THROUGH THE AZURE WATER AT about eight knots. I am kneeling on the bow, binoculars in hand, searching the horizon for turtles and dolphins. We are alone on the mirrorlike water, except for occasional flying fish and frigate birds.

Soon we are joined by two adult Atlantic bottle-nosed dolphins, surfing the wake on our port side. I make eye contact, and it seems as though they smile back. But I know better. Or do I? Suddenly three dolphins jump out of the water right in front of me. I lean over the bow for a picture, and they begin swimming in a crisscross pattern just below the surface, maintaining a speed that keeps them about three feet ahead of us.

I signal the captain to stop the boat, hoping to get underwater photos of the dolphins. He says they will leave as soon as I get into the water. I will take the chance.

To my delight, when I jump into the water with only my camera, mask, fins, and snorkel, the dolphins swim up to me and do headstands, making clicking and squeaking sounds.

The dolphins regroup and swim down to the sandy bottom at eighty feet, then turn suddenly and approach me at top speed. My heart is pounding and I am holding my breath. Just as I am about to close my eyes and brace myself for contact with these six-hundred-pound animals, they break into two groups and speed past me on my left and right.

When I return to the boat, I discover that in all the excitement I have forgotten to remove the lens cap on my camera. Since we have dropped anchor, I decide to make a second dive in scuba gear. I descend to the bottom and

During the day, parrotfishes break off small pieces of coral with their beaklike mouth to get at algae. At night, these herbivores sleep in the reef or on the sandy bottom, secreting mucous cocoons that help protect them from nocturnal predators. More than a dozen parrotfish species populate the Belize Barrier Reef, making identification a challenging task for the novice diver, especially since males and females have different colorations at different times of day and at different stages in their lives.

Facing Scientists don't always agree on the reason for a fish's coloration and pattern. Some believe the black band over the banded butterflyfish's eyes hides them so predators won't know which way the fish will flee when attacked. Others feel the series of bands works to confuse predators.

photograph the reef. Although I can't find the dolphins, I hear their clicks and squeaks throughout my thirty-minute dive. They remain somewhere nearby, but for some reason, they are careful not to approach a bubble-exuding scuba diver—something I have found to be true wherever I have dived with wild dolphins.

. . . .

FULL MOON, CALM SEAS, NO WIND, CLEAR WATER— CONDITIONS ARE ideal for a night dive.

I am standing on the dive platform checking my dive gear, especially my dive lights. It is 9:00 P.M., and I am ready to enter the fascinating world of Glover's Reef at night.

I look down into the blackness and see the reflections of several three-foot-long fish. Their dorsal fins are just barely breaking the surface, and their pectoral fins are outstretched in the hunting position. They are circling the dive platform in ever-tighter circles. But there is no need for alarm. These nighttime predators, the tarpon, feed on small fishes and are not easily approached by divers. They are stalking a large school of silversides that hover under the boat.

My dive buddy and I plunge into the black water. Turning on my dive light, I descend toward the bottom, which I cannot see. I am reminded of driving on a foggy road late at night, when my vision is limited to a narrow beam of light. Underwater, I feel more vulnerable.

Looking up toward the dive boat, I see the blinking strobe light hanging from the dive platform. This signal is our method of finding our way back to the boat, so we must keep it in view at all times. Of course we could surface and check the boat's position every so often, but word of sea wasps (a type of jellyfish whose sting can make a grown man cry) dissuades us from this navigation technique.

As we approach a depth of forty feet, our dive lights illuminate the spurs and grooves of the reef. Shining my light into a crevice, I find a parrotfish surrounded by its protective mucous cocoon, secreted after nightfall so predators, including moray eels, cannot detect its presence.

Nearby, two sleeping butterflyfish glide gently back and forth in the slight surge. I am surprised that even when they knock into the reef, they don't wake up.

*Most reef fishes are quick-change artists. The hogfish, **top left**, is usually orange during the day, blending in with the colorful reef. At night, **bottom left**, sleeping on white sand, it turns white, becoming less visible to nighttime predators.*

***Top right** When threatened, balloonfish inflate their bodies with water, expanding from the size of a large fist to the size of a soccer ball. Spines sticking out in all directions make the inflated fish an unappetizing meal choice for predators. Balloonfish have sharp teeth, capable of crushing the shells of mollusks and crabs— or the fingers of a diver.*

Even my camera strobe doesn't startle them.

Under a ledge, I find a four-inch-long balloonfish. Accidentally, I alarm the fish, and it immediately inflates with water, puffing its body to the size of a soccer ball, but one with dozens of two-inch-long spines protruding in all directions, a good defense against predators. The sharp spines could puncture my skin, so I swim away.

I see a sleeping hogfish and move in for a photograph. Like many sleeping fish, the hogfish changes color to blend with its surroundings. During the day, as it forages, the fish is a bright orange. At night, resting on the sand, it is almost entirely white.

On night dives, photography is my prime concern, but I also concentrate on two other things. First, since visibility is limited, I'm extremely careful of my movements so I don't damage delicate corals. Second, long-spine sea urchins live on the reef, waving their porcupine-type spines. Swimming too close to the reef, I could get a long spine through my wet suit and into my skin, where it would painfully remind me of my dive for a day or two.

Out of film, I follow the blinking strobe back to the dive boat, where I won't have to worry about being stung, impaled, or bitten by nocturnal sea creatures.

. . . .

TODAY WE WILL BE DIVING NEAR HALF MOON CAY ON LIGHTHOUSE Reef, twenty miles north of Glover's Reef. Half Moon Cay is Belize's oldest protected site; in 1928, while this country was a British colony, the government designated the area a Crown Reserve Bird Sanctuary, and in 1982, the newly independent government of Belize declared it a natural monument.

Several unusual sea birds enjoy protection at Half Moon Cay, including the red-footed booby and the magnificent frigate bird. About four thousand boobies nest each year in a forest of siricote trees, once the climax forest on most of Belize's cays and now found only here. Both boobies and frigate birds are drawn here by the atoll's abundance of reef fishes.

Lighthouse Reef is equally renowned for its forest of gorgonian corals at Northern Cay and for the Blue Hole, a sinkhole cave with stalactites and stalagmites from Pleistocene times.

The sun crept over the horizon an hour ago, and I am suited up at 6:00 A.M., my favorite time to dive and photograph, when the change between the day reef and night reef is quite evident, and the early light creates dramatic shadows on the reef. As nocturnal and diurnal creatures cross paths, I will see many fish feeding—fish eating fish, for the most part.

As we swim toward the reef, purple sea fans and yellow and green tube sponges are silhouetted against the rising sun. Occasionally, a school of blue chromis swims by, adding an element of movement to the scene. Suddenly, I see a sergeant major pass me at top speed, followed by a three-foot-long shark. In an instant, the sergeant major is gone, and the shark slowly swims on. I am thankful that this shark prefers three-inch-long fish to six-foot-long divers.

At sixty feet, the reef has a blue tint. This happens because water filters out color selectively, starting with the reds and ending with the greens and blues.

Sea fans in Belize, which reach heights of four feet or more, look like giant plants. In fact, they comprise thousands of individual coral animals which trap plankton in their tentacles.

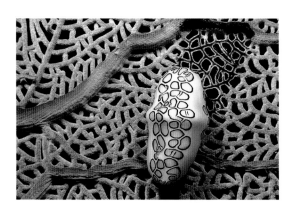

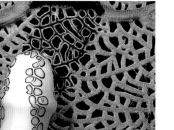

The flamingo tongue grazes on soft corals, leaving a dark path in its wake. Athough the inch-long animal is a mollusk, its beautifully patterned exterior is not its shell, but its mantle, a thin tissue which secretes the shell. When threatened, the flamingo tongue retracts its mantle inside its shell, giving collectors, who shouldn't be picking up live shells in the first place, a big disappointment.

However, when I turn on a dive light, the scene explodes with bright reds, yellows, greens, and oranges, and the camera's electronic flash enables me to capture the true colors of the reef inhabitants on film.

Before our time runs out, we swim to the reef edge and hover effortlessly over a thousand-foot drop-off. I wonder what strange and beautiful creatures lurk below in the darkness.

Breaking the surface, I am greeted by flat seas and a warm sun. I make a 360-degree turn and see only our dive boat and Half Moon Cay off in the distance. I enjoy the feeling of solitude, and hope that I can take this feeling back to New York with me.

Late in the afternoon, we dive again off Half Moon Cay. The sky is filled with huge white clouds, and dozens of frigate birds from the protected cay glide on air currents. The wind is picking up, as it often does in the afternoon.

Underwater, the sun now shines directly on the reef wall. A few hours earlier, the corals, fishes, and sponges had been in the shadows, hidden from the rising sun by the reef crest. Now I can clearly see what can only be described as wall-to-wall life. Virtually every inch of space is covered with some type of hard or soft coral, sponge, or limestone-secreting algae. A myriad of strangely shaped reef fishes glide up and down and back and forth on this vertical drop-off. I wonder who is watching whom?

Soon I come upon the entrance to a cavern at a depth of fifty feet. Entering the narrow passage, I am careful not to stir up the silt on the sandy bottom, which would drastically reduce visibility and make it very hard for me to exit safely.

Ahead of me is a coral-encrusted overhang. Deep in this crevice, I find a familiar sight, a nurse shark resting in the soft sand. Nurse sharks are commonly found resting under overhangs, and experience has taught me that these animals present no danger if left undisturbed. I observe the sleek brown animal for a few minutes and move on. Did it know I was there?

More than a dozen species of coral grow in this grotto—a bewildering variety to a novice diver. With a little reading and research, however, underwater explorers can identify them.

This dive site is called the Cathedral, getting its name from the cathedral-like spires rising from the reef. I find it an awe-inspiring place, and think, yes, this is like a cathedral or church, a place to reflect on life and come to peace with oneself.

. . . .

IN THE EVENING AFTER MY LAST DIVE, I THINK AGAIN ABOUT the future of Belize and its extraordinary barrier reef. I remember a recent conversation with a local fisherman in Placencia, a village on the southern coast, a hundred miles from where we are anchored tonight.

Villamar Godfrey shared some fascinating facts with me over Belikin beers. He said he used to be able to predict the weather by noting which way small lizards face in the evening. He can't do this today because changes in the weather pattern, caused by the destruction of the mangrove forests, have confused the lizards, and they are simply not reliable any more.

Villamar also told me about hungry boobies, which "hang" themselves in trees when they cannot find food rather than starve to death. He also said he believed that pelicans, which dive into the water every day for food, intentionally break their necks on their last dive when they are too old to fish any longer.

When we talked about conservation, Villamar expressed concern for the reef, which he feels is being hurt by the chemical run-off from the citrus farms on land and the dredging of the mangroves for deep-water harbors.

We ended our conversation on a positive note. "You want to dive with a whale shark?" he asked. "Just look for sea birds a mile or so from land. You'll find the whale shark there."

The concerns of fishermen like Villamar are taken very seriously by Jacque Carter, a professor of biology at the University of New England. Because of his familiarity with conservation problems in Belize and around the Caribbean, he was invited by CEDAM International to its initial Seven Wonders meeting, where he made the case for including the barrier reef. Later, when I told him about my plans for the book, he sent me an entertaining account of his contribution to the creation of a marine reserve in Belize.

The British colonial government established Belize's oldest nature preserve on Half Moon Cay at Lighthouse Reef in 1928. Among the sea birds which enjoy protection here are the red-footed booby and the magnificent frigate bird.

Carter went to Belize in 1984 under the auspices of Wildlife Conservation International to help develop a research and conservation program for the barrier reef. He had heard disturbing reports from government officials and local fishermen about declines in catches of Nassau grouper, a mainstay of the local fishery.

Carter, his wife, and their two-year-old son settled into a beach house facing the reef, and he began his fishery studies. "But as I struggled endlessly with leaky boats and wheezy motors," he writes, "it became apparent that I needed help."

One morning, after a breakfast of tortillas, fried plantain, and refried beans, Carter kissed his wife farewell and ventured forth to find a fisherman who would be willing to work with him. "Being a learned man," he reports, "I knew that fishermen the world over could be found in one of two places: at sea, deftly handling baskets of cold fish, or in port, lashed to some bar stool, handling bottles of beer with equal aplomb."

His wanderings led him to Fido's, the island's piano bar, perched atop salt-soaked timbers overlooking the reef. Here an old fellow named Bill sometimes plays requests on an out-of-tune upright that shares a corner with a dart board.

On this morning, three shirtless fishermen were at Fido's, holding down a table near the door. The largest of the three had a screaming eagle tattooed across a chest that was easily the size of a fifty-five-gallon oil drum.

Palms sweating, Carter stepped forward and asked in his best Spanish if any of these men would be interested in helping him with his research.

They all stared with the greatest seriousness for what seemed like an eternity at the sober, bespectacled gringo standing nervously before them. Then the tattooed one broke out in a smile that would rival the sun and invited him to join them for a drink.

As Carter pulled up a chair, the waiter appeared with six cold beers. He placed one in front of each fisherman and three in front of Carter. Sensing the American's confusion, his newfound compadres carefully explained that each of them was honored to buy him a beer. He had no choice but to imbibe or risk offending their generosity and goodwill.

The Nassau grouper can change color from pale to almost black. But its transformation skills don't stop there. All groupers are born female; some change to males as they mature.

At night (when this photograph was taken), sharptail eels leave the reef to forage in the sand and on patch reefs. Novice divers often mistake harmless sharptails for poisonous sea snakes, which do not inhabit the Caribbean.

"The hours passed," Carter recalls, "and I became proficient at opening beer bottles with a fillet knife. The big smiling fisherman had no need of such conveniences. He just used his teeth. At nightfall a worried American wife was greeting three singing fishermen supporting one staggering American ichthyologist at her front door."

The next morning, the fishermen were again at Carter's door. But this time, they were ready to work. Later that year, the one with the eagle tattoo would save the life of Carter's field assistant as she struggled with scuba gear in deep water, out of air and unable to fight strong currents and heavy seas. "We all became very close friends," says Carter.

As time passed, they began to learn about Nassau grouper and the reef on which it depends. They outfitted individual fish with ultrasonic transmitters to chart grouper home ranges, movements during feeding forays, and interactions with other fishes. Late each afternoon, they would meet at the water's edge where the local fishermen cleaned the day's catch. On sun-bleached paper, they scribbled down the weight and length of hundreds of fish destined for market. They inspected each fish to determine its sex and reproductive condition and examined its scales for rings, called annuli, that reveal age and growth rates. And they logged many hours underwater documenting the behavior of these curious creatures.

"Our field work indicated that the grouper fishery was headed for trouble," Carter writes. "Spear fishing by sport divers, combined with decades of commercial fishing on the banks, was causing a decline in stocks. If these trends continued unchecked, the fishery could collapse, as it had elsewhere in the Caribbean."

Carter needed several years of research to propose a complete management model. But he feared that the grouper fishery might collapse unless action was taken sooner. Armed only with preliminary results, he began work on a plan for the long-term protection of Nassau groupers and their habitat.

Identifying the habitat boundaries of these highly mobile fishes and responding with plans for protected areas was a major problem. In theory, scientists strive to strike a balance within reserve boundaries between the rate of species loss to the system and the rate of species replacement. This balance is best achieved in large protected areas because portions of the reef damaged by natural and human causes can be replenished by colonists from undamaged parts of the same reef.

Unfortunately, financial, political, and practical constraints often make it impossible to protect large reef areas. In Belize, Carter proposed establishing smaller parks along the barrier reef which could eventually be connected in a necklace of protected marine habitats.

The plan had to win support from the fishermen who make their living in these waters. Experience has shown that protected areas and species management plans based solely on scientific criteria stand little chance of success. Wildlife Conservation International worked closely with the government of Belize to establish a reserve system that incorporates traditional customs in its management plans.

On May 2, 1987, after three years of study and months of often-heated debate among the scientists, fishermen, tourist guides, and others who care about the reef, legislation was signed officially establishing Belize's first marine reserve at Hol Chan, four miles from San Pedro. Hol Chan's centerpiece is a striking natural break in the reef where nutrient-laden lagoon and ocean currents give rise to unusual coral formations and a variety of fishes and other sea life, including Nassau grouper. In all, the reserve encompasses over five square miles of connected coral reef, sea grass meadow, and coastal mangrove habitats.

Hol Chan is remarkable not only for its breathtaking natural splendor, but for the management issues it seeks to solve. For Belizeans, who are prepared to take responsibility for this underwater wonder, development and conservation are not necessarily destined to be conflicting goals.

Encouraged by the success of Hol Chan, Carter is cautiously optimistic that the fledgling marine reserve system will grow stronger with time. Thanks to the conservation efforts and programs of the Belize Audubon Society, the Belize Department of Fisheries, the United States Administration for Internal Develop-

The sand-and-seagrass beds of the lagoon at Hol Chan Marine Reserve serve as a juvenile staging area for many species of fishes and crustaceans. Grunts, snappers, surgeonfishes, and parrotfishes also graze on algae and seagrass here, and the lucky diver may even see a turtle or a manatee.

Elkhorn coral flourishes near reef crests amid pounding waves. But strong as it may seem, this common Caribbean species cannot withstand carelessly dropped anchors. With a growth rate of about an inch a year, this colony will take thirty-six years to replace a three-foot-long broken branch.

Facing *Throughout the 160-mile-long Belize Barrier Reef ecosystem, marine explorers will find solitary coral heads, underwater oases that attract dozens of different species of fishes, corals, and invertebrates.*

ment, CEDAM International, Wildlife Conservation International, and the New York Zoological Society, more plans for the reef are taking shape. Marine parks have been established where spear fishing, turtle hunting, and coral collecting are illegal, and more protected areas are being proposed. Eventually a Belize Barrier Reef Authority will supervise the protection, sustainable use, and recreational enjoyment of the barrier reef through a zoning system, designating areas suitable for fishing and research and areas that must be strictly protected to preserve genetic diversity and maintain essential ecological processes.

More good news for conservationists is that most of the dive boat operators, like the crew of the *Belize Aggressor*, are conservation minded and brief their divers on the fragility of coral reefs. Permanent mooring buoys, which protect the reef from anchor damage, have been placed at several popular dive sites. Over the next several years, additional "Reef Savers" will be installed.

These things are all good for Belize's natural resources. But what would the outlook be for this underwater wonder without the environmental concerns of Belizeans and the international organizations? One does not have to look far to find the answer.

Toward the northern end of the reef, off the coast of the Yucatan Peninsula, mangrove destruction and road and hotel construction have resulted in many silt-smothered, dying, and dead reefs. Overfishing in this area, as well as extensive turtle collecting, has also shattered the ecological balance. In fact, on my last dive off the Yucatan in 1988, I was hard pressed to see any large fishes on the reef.

Near the southern border of Belize, the reefs are beginning to show signs of damage. Fishermen from other countries which have overfished their own areas are now fishing these waters. If their unregulated harvests continue or increase, we cannot hope for a thriving coral reef community in the years to come.

George Page, host of PBS's "Nature" television series, put the situation quite frankly: "We are fortunate to be the first generation ever to witness the beauty of the kingdom beneath the waves. It would be an unconscionable sorrow if we were also to be the last."

LAKE BAIKAL

ON A JUNE DAY IN 1990, SUSAN AND I ARE WALKING ALONG THE SHORE of Lake Baikal in Siberia. We've been here three days, and although we are eager to begin diving, we are still too affected by jet lag to work safely in Baikal's icy water. There's a thirteen-hour time difference between our home near New York and this lake twenty-five hundred miles east of Moscow; for us, a complete reversal of night and day.

Even if we had to travel farther, Lake Baikal would be worth it. Among our seven underwater wonders, Baikal stands out. It is our only freshwater site; all the rest are salt water marine environments. It is our only continental site, ringed by mountains, twelve hundred miles from the nearest ocean. And it is certainly the coldest place that Susan and I will be diving; our other five dive sites are all in the tropics, and the deep ocean vents are miles below the reach of any diver.

In a way, Baikal is serving as a representative in our project for the wonders of all the planet's freshwater environments. And what a representative! Geologists estimate that Baikal is twenty-five million years old—the oldest lake on earth. Lake Tanganyika, the second oldest, is also centered on a rift, is almost as deep, and may be twenty million years old. The Great Lakes, by comparison, go back a mere twenty thousand.

One-fifth of the world's fresh water, excluding what's frozen in the ice fields of Antarctica, is contained in Baikal's nearly four-hundred-mile-long expanse. To empty Baikal, the lake's sole outlet, the Angara River, would have to flow at its present rate for four hundred years. No other lake on earth contains so much fresh water; even the five Great Lakes combined contain less water than Lake Baikal.

In summer's calm conditions, freighters travel back and forth between cities and towns around Lake Baikal. In January, the coldest month, only the southern part of the lake near the Angara River does not freeze. A three-foot-thick sheet of ice covers the rest, and trucks, motorcycles, and sleds are the only ways to get across. During the winter, scientists drive far out onto the lake to study Baikal seals in their ice dens.

Facing *Baikal's chilly waters are home to a group of bottom-dwelling fish species called sculpins, which spend most of the day and night lying motionless on green sponges. Because sculpins live on the lake bottom, silt and sediment from logging threaten their survival.*

Baikal's summers are short and relatively warm, reminiscent of early fall in the northeastern United States. The lake's water temperature, however, peaks at a chilly 38° F, only two degrees warmer than in January. The cold water creates a microclimate along the shore where vegetation typical of more northern regions predominates.

Yet among the world's largest lakes, Baikal ranks only eighth or ninth in surface area. The immense volume of water in Baikal, perhaps five thousand cubic *miles* of water, is accounted for by its depth, 5,380 feet at its deepest. That's not only the world's deepest lake bottom, but the lowest point in the planet's continental surface.

What's more, despite the influx of sediment from over three hundred rivers and streams, Baikal isn't getting any shallower. The lake is situated on a continental rift, a place where the earth's tectonic plates are pulling away from each other. Over the millennia, the deepening of the rift has kept pace with sedimentary accumulation, and in places, Baikal's sediments are estimated to be over three miles deep!

While we are diving amid Baikal's shallow-water wonders, Emory Kristof, a National Geographic staff photographer, and Dr. Kathleen Crane of Columbia University's Lamont-Doherty Geological Observatory will explore its depths, using a *Pisces* manned submersible vessel and ROVs (Remotely Operated Vehicles) to examine geothermal vents on the lake bottom. We are excited that our expedition coincides with theirs and are eager to hear the results of their work.

. . . .

SUSAN AND I SPEND THE DAY PHOTOGRAPHING THE COUNTRYSIDE, which reminds us of upstate New York in spring. The day is windless, and the surface of Lake Baikal reflects the white clouds and blue sky of summer. As we walk amid tall pine and birch trees by the shore, we hear many songbirds. We see wildflowers everywhere, north Asian ones, unknown to us, their whites, yellows, oranges, and purples lining the winding path. We've read that bear, moose, sable, and fox live in these woods, but the largest animal we encounter is the famous Baikal blue squirrel.

Although we can see snow-capped mountains on the opposite shore, thirty miles away, we find it hard to believe that we're in Siberia, where February temperatures drop to -40° F, the lake surface freezes in an ice sheet three feet thick, and the wind blows hard enough to lift the roof off your house—if your house isn't already up to the rooftop in snow.

We soon discover how quickly even summer weather turns Siberian here, when a dense fog rolls off the mountains, shrouding the lake in a motionless grey veil. Then rain and wind begin, thoroughly chilling the air. The storm passes as quickly as it begins, though, and before long the sun again warms our faces.

When we return to our base at the Limnological Institute in the picturesque village of Listvyanka, our hosts tell us that Baikal is famous for these sudden tempests, particularly in the autumn. Siberians have special names for two

Unlike the waters of other very deep lakes, the waters of Baikal contain oxygen at every depth, even near the bottom, and fish inhabit the entire water column. Lake Tanganyika, the world's second deepest lake, supports fish only to a depth of 650 feet, below which there is no oxygen.

of the lake's more menacing winds. The *barguzin*, which blows straight down the lake for four hundred miles, causes big, oceanlike waves. The *sarma*, a more vicious but less sustained gale, produces gusts of up to eighty miles per hour, creating a ferocious chop that's deadly to small craft—like the ones we'll be using.

It's conditions like these that explain the Limnological Institute's policy of overseeing all scuba dives in the lake. As guests of the Institute, which is part of the Siberian Branch of the Soviet Academy of Sciences, we feel very well cared for. Everyone here, from the director, Dr. Mikhail Grachev, to the team of lake scientists, headed by Dr. Vadim Fialkov, is going out of his or her way to make our fourteen-day expedition a success.

As I think about the Institute and its long history as a research center, going back to 1918, I recall what Dr. Andy Rechnitzer, a member of our expedition, has told me about the unique flora and fauna of Baikal. With over fifteen hundred animal species and a thousand plant species, Baikal is much richer than most lakes in its variety of living things. Biologists attribute this richness to high levels of oxygen at all depths in the lake, not simply near the surface, where oxygen is normally concentrated. They believe that Baikal's oxygen-rich water circulates from the surface to the bottom, a pattern possibly set in motion by deep geothermal vents.

What is more, many of Lake Baikal's species—over two-thirds—are found nowhere else in the world. Situated in mid-continent, the lake has been isolated from other aquatic environments for much of its history. The species which have prospered here have often diverged sharply from their relatives elsewhere. For modern researchers, the lake is like a vast laboratory of evolutionary history. "An understanding of the evolution of species in Baikal," Andy tells me, "will help us develop better models for evolutionary processes worldwide."

· · · · · ·

ON THE FIRST DAY OF DIVING, WE ARE FILLED WITH ENTHUSIASM FOR exploring the relatively unknown underwater region of the lake. We also feel a sense of urgency; our "window" for photography is fairly small. At this time of

Tall sponges are common on coral reefs, but not in freshwater lakes. Baikal sponges, of which there are no less than ten species, get their green color from zoochlorella, the microscopic algae that live symbiotically in the sponge's white tissues. This species of branching sponge grows abundantly to a depth of sixty feet near the lake's southern end.

year, underwater visibility in Lake Baikal can range up to 150 feet. Even from the surface, we can clearly see rocks and sponges fifty feet down. But soon, as the water warms, the lake's plankton will begin to bloom, reducing visibility to less than ten feet.

Our first dive in Baikal is as different from a tropical reef dive as night is from day. In the tropics, color, action, and variety are visual stimulants on every dive. In Baikal, everything is tempered with a soft green tint. The mood is serene; animals just don't move fast in cold water.

The most striking difference is the water temperature. In the tropics, we're comfortable diving in our wet suits in 84° F water. Here, the water is 38° F. When we enter, it feels as if we've been hit in the face with a giant snowball. The sensation does not go away but increases in intensity. Although our bodies stay fairly warm, thanks to our dry suits and thermal underwear, our fingers become so cold that it's hard to work the camera and suit buoyancy controls. The cold water in our ears affects our equilibrium, so we're often dizzy, especially when we explore the upper faces of the lake's huge underwater cliffs.

We're used to diving for an hour or more at a time on tropical reefs, but here the cold water limits us to twenty minutes. That doesn't leave much time to be choosy or make mistakes, so we try to get the most out of every dive and every roll of film.

After experiencing the chill of a June dive in Baikal, we feel only admiration for our hosts, the Russian scientist/divers, who cut holes in the thick ice in midwinter and dive below, searching for specimens in water near the freezing point—in wet suits, no less. But more on this later.

.   .   .   .

MOST OF OUR DIVES ARE JUST OFFSHORE NEAR LISTVYANKA AND THE Institute. We explore the massive boulder and rock formations that line the underwater lake slope. Many are covered with green, fingerlike sponges—*Lubomirskia baicalensis*, the trademark of underwater life in the shallows of Baikal. These sponges, the most impressive of the lake's ten sponge species, get their green color from *zoochlorella*, a type of algae which lives symbiotically in the sponge's tissues.

Lake Baikal is known for its exceptional clarity in early summer. Forty feet below the surface, scuba divers can see the outlines of clouds in the sky. Exploring the frigid underwater environment requires a dry suit, which keeps cold water out and body temperature in—if the suit doesn't leak. Still, dives are often limited to only twenty minutes, after which hands and faces start to freeze.

*Aquatic creatures all over the world use camouflage to avoid predators. This species of Baikal amphipod fades into its surroundings when walking on algae-covered rocks, **top**, but becomes vulnerable when grazing on green sponges or swimming in open water, **bottom**. Eighty percent of the world's known amphipod species are found in Lake Baikal.*

Near the surface, *Lubomirskia* grows in barklike patterns, covering the stony bottom with a bright green carpet. The sponge changes its character at depths of ten to twelve feet, putting out branches from a broad base. At their largest, between depths of fifteen and forty feet, these branches reach a foot or two in length. Intertwining and connecting with neighboring sponges, they make an almost reeflike structure that spreads down rocky slopes to points eighty or a hundred feet below the surface.

Like a reef's convoluted shapes, these sponge groves are an ideal habitat for smaller aquatic creatures. Our deeper dives near the groves, below forty feet, take us into the habitat of Baikal's gammarid amphipods. "This class of crustaceans," Andy Rechnitzer tells us, "has expanded in Baikal to fill the niches held by shrimp in other underwater sites."

Baikal's amphipods, represented by no less than 240 species, are bottom-dwelling scavengers bearing a striking resemblance to terrestrial insects, complete with tentacles, hairy legs, and segmented bodies. One small species, the inch-long *Axelboechia carpenteri*, is as purple as grape juice. Others are covered in protective armor; *Acanthogammarus maximus*, for example, bears large spines on its carapace.

Encrusting sponges are also widespread, forming extensive green aquatic carpets. We notice several snail species, ranging in size from a dot to a dime, creeping over these emerald surfaces in search of a meal.

Green algae cover the rocky lake bottom; Baikal is very rich in algae, with over 570 species. In the shallows, we swim past lettuce algae, smoke algae, and hemplike algae. As we move through the water, it's nearly impossible not to disturb the algae meadows. Even the slightest motion creates dense green clouds around us.

The all-pervasive green of this chilly underwater kingdom is rounded out by Baikal's epiphytes, which cover large portions of the bottom, spilling like miniature waterfalls from rocks and other surfaces.

Exploring these algae meadows and sponge gardens is like being in a dream. We'd seen no underwater pictures of Baikal before we arrived, and visually, it's a far stranger and more beautiful scene than we could have imagined.

The clarity of Lake Baikal's water is legendary, and now that we're underwater, the distances we can see continually amaze us. This clarity is due to the water's relative lack of minerals, and also to the presence of a very small crustacean, *Baikal epishura*, which consumes large amounts of the algae, plankton, and bacteria which normally cloud fresh water. The word "pristine" takes on new meaning when applied to a space four hundred miles long and up to a mile deep, filled with almost perfectly clear water. There is no place like it in the world.

Vast algae-and-sponge gardens have given Baikal the nickname "the green reef." In early summer, when this photograph was taken, visibility can reach more than a hundred feet, better than in most lakes and equal to many of the world's coral reefs. By midsummer, the algae blooms and visibility dwindles to about ten feet.

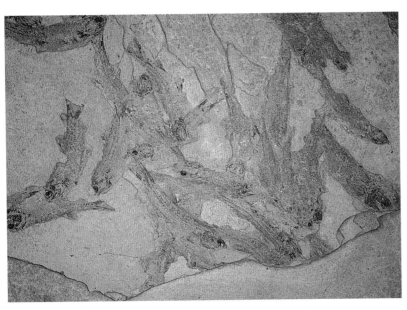

Fossilized remains of prehistoric fish offer scientists a glimpse of Baikal life twenty-five million years ago when the lake was forming. Baikal is hundreds of times older than all but a few of the world's lakes. Most date back only to the end of the most recent ice age some ten thousand years ago.

Because Baikal is so deep, so clear, and so alive with unique species, the destructive effects of industrial development at the lake's southern end seem particularly brutal and thoughtless. Here, effluents from the huge Baikalsk mill, built to manufacture rayon cord for airplane tires from wood pulp, have severely contaminated many square miles of the lake. Environmentalist and Baikal expert John Massey Stewart reports that the Baikalsk plant releases the same quantity of effluent per day as the city of Los Angeles.

What's worse, the Baikalsk plant was built to make a product which was outdated within a few years of the plant's completion in the early 1960s. And even if the plant had been necessary, it could just as well have been built downstream on the Angara River. But despite the weak market for Baikalsk tire cord, and despite thirty years of opposition from Siberian environmentalists, Baikalsk has continued to operate—partly because it provides thirty thousand Siberians with jobs. Purifying systems more expensive than the plant itself have been installed, and part of the complex now manufactures paper rather than rayon cord, but the air and water of Baikal continue to be fouled by Baikalsk's industrial waste.

Things aren't much better at the lake's northern end, where logging has cleared vast areas of the taiga forest, causing erosion that carries life-choking sediment into the surrounding rivers and toward the lake itself. In the short Siberian growing season, new forests grow slowly; decades will pass before the water-filtering effect of the mature taiga forest is restored in this part of the Baikal watershed.

The industrial threat to Lake Baikal goes far beyond the notorious Baikalsk pulp plant and the clearing of the taiga. Altogether, there are more than one hundred factories on Baikal's shores and tributaries, especially near the city of Ulan Ude on the Selenga River. Few of these factories include any effective pollution-control devices. Inadequately treated sewage from Ulan Ude and other growing cities and towns around Baikal is also a problem. Rich in nitrates, sewage stimulates certain kinds of algal growth. In the delicate biological balance of Lake Baikal, a sudden explosion in some algae populations would not only cloud the water, but set off changes throughout the food chain.

Evidence is mounting that Baikal's ecological equilibrium is already on the verge of collapse. Omul, a delicious fish endemic to Baikal which has long been a major food in Russia, is now spawned at hatcheries and released into the lake because it no longer reproduces naturally. Many other fish species are said to reach only half their normal size. Populations of *epishura*, whose filtering action is so important to the lake's water quality, have begun to drop. And in 1987, thousands of Baikal seals, the world's only freshwater seal species, died in an epidemic. Russian scientists agree on the cause of death: the lake's polluted water, which has compromised the animals' immune systems. The epidemic was both a clear indication of how severely the Baikal ecosystem has already been affected by industrial development and an important new factor in the lake's continuing deterioration.

As this "Save Baikal" sign illustrates, local conservation groups are fighting to protect Siberia's "Sacred Lake." Unfortunately, two paper pulp plants near the lake's southern end have been dumping toxic effluents into Baikal since the early 1960s. Siberian, Russian, and international environmental groups hope to convince the plants' operators to halt this destructive practice. In 1990, UNESCO nominated Lake Baikal a World Heritage Site. A WHS designation would encourage coordinated interdisciplinary and international scientific research into the lake's unique ecosystem.

There are signs, however, that the thirty-year struggle of Siberia's increasingly vocal environmentalists has begun to influence the future of Lake Baikal. In fact, the movement was the Soviet Union's first environmental campaign, and it gathers some of its authority from the belief among many Siberians and Russians that Baikal is an inviolable symbol of the motherland. Through the efforts of these environmentalists, some progress has been made in controlling industrial pollution, but far more remains to be done.

Early in Mikhail Gorbachev's administration, several extensive stretches of Baikal's shoreline became national forest parks, with a three-kilometer zone in the lake where regulations forbid environmentally harmful activities. These areas have not been declared sanctuaries, underwater parks, or aquatic reserves, but they are patrolled by forestry officers. We can only hope that in time, as the Siberian environmental movement gathers momentum and international awareness grows, stronger protective measures will be enacted—especially if, as UNESCO has recommended, Lake Baikal and its watershed become a World Heritage Site, like another underwater wonder, the Galápagos Islands.

WE DON'T ENCOUNTER ANYTHING LIKE THE VARIETY OF FISHES WE'RE used to in the tropics. But on every dive, we see forty to fifty sculpins, ancient bottom-dwellers ranging from four to seven inches in length. Lake Baikal is home to about forty sculpin species—an astonishing eighty percent of the lake's biomass of fishes. Among these, there are two, Andy tells us, which have the unusual ability to live anywhere throughout the lake's water column, from its maximum depths of over a mile to just beneath the surface.

The sculpins are unafraid of us, and they are easy to approach and photograph. Other fishes, including whitefish, salmon, sturgeon, and Siberian perch, are in the area, but they elude my camera.

Baikal does have something in common with the tropics—spectacular underwater slopes, continuations of the steep mountains that abut this rift lake's shorelines. In some places, the sheer vertical drop-offs plunge several thousand feet. We carefully descend into the dark green abyss. The deeper we go, the colder and darker it gets.

Investigating an underwater precipice, I lift a stone in search of a photographic subject. Suddenly, the surrounding rocks give way, and an underwater avalanche begins. I'm engulfed by a cloud of sediment, which reduces visibility from sixty feet to about two feet. I turn and swim toward open water, check my buoyancy, and follow my bubbles to the safety of the shallows. Lesson learned: look but don't touch in unfamiliar territory.

THE DAYS PASS QUICKLY. WE STILL DON'T EXACTLY ENJOY THE COLD water, but we have grown used to three dives a day. And besides, we've heard a local tradition which makes it all seem worthwhile. It is said that if you put your hand in the lake, you add one year to your life; your feet, two years. If you are brave enough to swim in the lake, you increase your life by twenty-five years.

If this is true, after diving in Baikal for two weeks, we should live a long,

Above and facing More than fifty different species of fish are found in Lake Baikal. Sculpins predominate, comprising 80 percent of the lake's biomass of fish.

Sculpins do not release eggs and sperm like most fishes, but give birth to fry, or immature fishes. The spawnings are massive, as each female releases two to three thousand individuals.

Remotely Operated Vehicle (ROV) cameras focus on sponges near a geothermal vent discovered on the muddy bottom of Lake Baikal. This photograph was taken in 1990 during a major scientific expedition to Baikal sponsored by the National Geographic Society and the Soviet Academy of Sciences. (Photograph by Emory Kristof © National Geographic Society)

Right *Pisces, a Canadian-built submersible, begins its descent to the depths of Lake Baikal during the expedition. (Photograph by Emory Kristof © National Geographic Society)*

long time. Time to return, time to experience more of the wonders of what many of the local people call "the Sacred Lake."

Toward the end of our stay, we hear from Emory, Kathy, and the National Geographic team, who have had great success in their deep-water researches. At a depth of 1,350 feet, near the lake's northeastern shore, expedition members have discovered a geothermal vent field that supports a unique community of sponges, bacterial mats, worms, snails, and fishes.

This is a real find, as it is the first time such a community has been photographed in fresh water. Documenting the vent field helps confirm that Lake Baikal, like the Red Sea and Africa's Rift Valley, is a place where continental masses are being pushed apart.

"The vent area goes on and on," Emory tells me. "The communities of life resemble organisms normally found in an ocean, which gives weight to the theory that Baikal is an ocean in the making." Millions of years from now, Lake Baikal may resemble the Red Sea, a salt-water wedge between two continental masses.

EIGHT MONTHS HAVE PASSED, AND I HAVE RETURNED TO LAKE BAIKAL
to dive and take photographs under the lake's winter covering of ice. Susan isn't
with me this time; she is back in New York State, six months pregnant with our
first child. Although I'm with friends, I feel very alone and could use her moral
support. A trunk containing all my warm clothes did not arrive with my flight
from New York to Moscow. I'll have to borrow what I can. There are no winter
clothes for sale in the department stores here.

For nearly two and a half hours, our caravan has been mov-
ing steadily north through a white desert. Two Russians lead the
way on a Russian-made motorcycle, circa 1950, complete with sidecar
(which I hear is filled with bottles of vodka to keep the expeditioners
warm). Our team of three American divers and twelve Russian scientists and divers follows in three four-ton, heavy-duty trucks—one
for diving gear, one for living and sleeping, and one for dining.

I am in the front seat of the lead truck. As we move over
the frozen terrain, I wonder what Susan would think about this
adventure to beat all adventures: driving on the ice straight up the
middle of Lake Baikal.

Suddenly Victor, the motorcycle driver, raises his hand, and
the caravan comes to a skidding stop. Everyone jumps out of the trucks for a
football-style huddle. I don't speak Russian, but understand through our trans-
lator, Olga, that Victor has spotted a mile-long crack in the ice blocking our path
north. I begin to get nervous when I see Victor making the unmistakable hand
motion of a truck going under.

*In winter, four-ton trucks replace research vessels as the
main mode of transportation on Lake Baikal. As spring
approaches, cracks in the thawing three-foot-thick ice
can stretch for a mile or more, compelling researchers to
stop and weigh the dangers of driving over them.*

Despite my apprehension at resting on the three-foot-thick ice, I'm overwhelmed at the beauty of the frozen lake. We are twelve miles from either shore
and surrounded by breathtaking snow-and-ice formations. In places, the ice is
so clear that I can plainly see the water beneath.

While the scientists decide what to do, I walk away from the caravan to
photograph the scenery. I try not to think about how deep the lake is here—
almost five thousand feet.

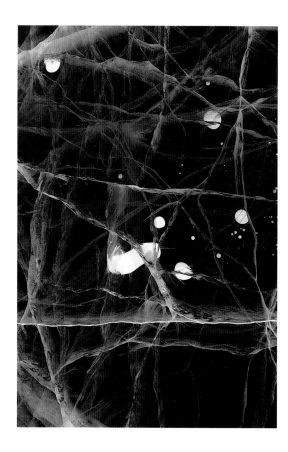

Locating seal dens on the frozen surface of a four-hundred-mile-long lake is easy if you know what to look for: bubbles that form under the ice when the seals exhale.

Right *Baikal seals are rarely seen on the ice at the southern end of the lake. This small pup may be looking for its parents. The Baikal seal's large eyes help detect prey in the darkness of deep ice-shadowed water. Its sharp claws help to grip fish–and to carve ice dens when the seal matures. A thick layer of fur keeps these mammals, the only freshwater seals in the world, warm in 36° F water.*

As I photograph, the ice moans, groans, and cracks. It's an unsettling feeling, but one I get used to. After all, the scientists have been exploring the lake in this manner for years, right?

When I return to the trucks, Vadim, the chief on-site Russian scientist, briefs me on how we will get across the crack. "Basically, the technique is quite simple," he says. "We find a relatively strong section of the crack and then drive the trucks over it as fast as they can go."

Okay. I'm willing to go along with this lake expert's plan. However, I take all of my camera equipment out of the truck, just in case.

As planned, all three trucks make it over the crack. Victor gives the "let's go" signal and we're on our way again.

Quite frankly, I never thought I'd be back in Baikal so soon after my trip in the summer of 1990. However, I wanted photographs that truly represent the lake, which is covered with ice almost half the year, from late November to early April. I also wanted to get at least one underwater picture of the elusive Baikal seal, the only aquatic mammal in Lake Baikal.

Baikal seals are big animals, reaching nine feet from the tip of the nose to the end of the hind flippers, and weighing from a hundred to nearly three hundred pounds. They are thought to have originally reached Baikal from the Arctic Ocean through the Yenisei-Angara river system during the ice ages.

At last count, sixty thousand of these silvery creatures were living in the lake, especially around the islands of the central and northern sections, where they sometimes bask on marble outcrops in fine weather. Because they are the largest species in the lake, the pinnacle of its food chain, the epidemic of 1987 caused a great deal of concern among people who study aquatic life in Baikal. If my pictures can help bring these seals and their wonderful habitat to the world's attention, this chilling northward trek over the ice, cracks and all, will be worth it.

. . . .

LAST NIGHT, WE MADE CAMP ON THE ICE AT AN EQUAL DISTANCE from the east and west shores of Baikal, about three hours north of the Limnological Institute. We chose this site for its proximity to a seal den that Victor had located before our arrival in Baikal.

It's now 6:00 A.M., and I'm ready to go diving. However, I haven't been out of our heated truck. Outside, it's 18° F, not taking into consideration the wind-chill factor, which drops the air temperature down below zero. Vadim says we must wait until it warms up. Maybe we can make our first dive after lunch.

Around 11:00 A.M., the Soviet divers drill a small hole in the ice about thirty yards from the seal den. Then they insert a manual saw and start cutting a larger entry hole. After about an hour, the hole is completed. But there is a problem: it's so cold that ice continually forms over the opening. The solution is to rake away the newly formed ice constantly.

While all of this is going on, I'm putting on my dry suit in the heated diving truck. Here, I'm warm as toast. However, that's all about to change.

Stepping outside isn't so bad—at first. By the time I reach the diving hole, my face, hands, and feet are cold. I have to work fast now, getting the rest of my diving gear on before I get too cold to go diving and before the diving gear freezes.

I'm all dressed, and I spit in my mask (as divers do before a dive) to prevent it from fogging. Sitting on the ice, my feet in the icy water, I rinse the mask. Before I get it on my head, ice forms on the faceplate. So much for that

Vladimir Batakov, chief diver at the Limnological Institute, tends a safety line for author Rick Sammon near a seal den. The three-foot-wide hole is the only way out from beneath the ice.

technique. Vadim gets a thermos filled with hot water, and I rinse the mask again. This time, it doesn't freeze.

I jump into the water, a lifeline securely tied to my harness. But I can't breathe through my regulator. It's frozen! Back to the surface I go for another dose of hot water. Finally, all the gear is working properly, and I begin my exploration of the eerie world under the ice.

Everything is upside-down—all solid objects are above my head. I am engulfed by green water, tinted by microscopic algae. Below me, there is only a black abyss. That is where the Baikal seals go for their meals, diving to depths of five hundred feet or more to dine on omul and other fishes.

It's a strange feeling diving under a solid layer of ice. I feel very, very alone. The only way out is through the solitary ice hole. With this in mind, I constantly check to see that my lifeline remains taut, held securely by a line tender on the surface.

The visibility is surprisingly good below the ice. In fact, the sun shines through brightly, providing me with about thirty-foot visibility. Slowly, I make my way toward the den, hoping beyond hope that I'll see a seal.

Looking up, I suddenly find myself in a hollowed-out area of much thinner ice. I'm in the seal's den. I rise to just below the surface and find long tunnels and dramatic ledges branching out from a central open area. This is where a pair of Baikal seals—which mate for life—have carved their home out of the ice while it was forming in late autumn. Here they will live for several months, and here they will raise their young.

After photographing the den, I start to make my way back to the ice hole. Turning, I can't believe what I see—a seal pup, staring at me from a distance of about ten feet. Thank God I have my camera ready and set on automatic. I quickly click off a shot before the seal darts down into the blackness.

We will make four more dives on this trip to Lake Baikal, and by the end of my stay, I've long since ceased to doubt the wisdom of making this risky winter journey. In fact, I would like to stay longer, but time is running out—the lake is beginning to thaw.

During our return trip to the Institute, I ask Vadim when they stop ice diving. Casually holding back a smile, he replies, "When the first truck goes under."

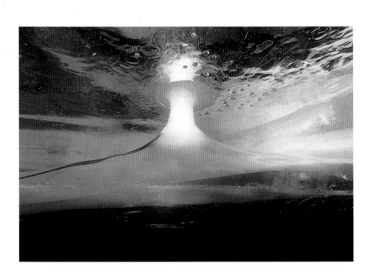

A column of ice supports the domed ceiling of a Baikal seal den. Beyond the column are tunnels where the seals, which mate for life, raise their young. A five-foot-wide oval opening in the floor leads to the lake water below.

Facing *Apparently unafraid of visitors, a one-month-old seal pup swims just below its den, the dark circle in the center. Adult seals carve the dens while the ice forms in early winter. Both parents and pup dive to depths of more than five hundred feet for food—omul salmon, perch, pike, and sturgeon.*

THE NORTHERN RED SEA

WE ARE ANCHORED OFFSHORE NEAR RAS MUHAMMAD, THE SOUTH-ernmost tip of the Sinai Peninsula in the northern Red Sea, where the Gulf of Aqaba and the Gulf of Suez meet. From the aft deck of the *Colona II*, our dive boat for the week, the topside scene is awe-inspiring, yet bleak. Barren mountains rise out of a monochromatic desert. The wind is un-usually strong for September, and dust from a sandstorm blocks the warmth of the late afternoon sun. Through the haze, the sun is a glow-ing white ball—a perfect sphere—suspended in a cloudless grey sky. Our Norwegian skipper, Captain Freddy Storheil, says he hasn't seen anything like it in all his years of running charters on the Red Sea.

The area looks lifeless; only a few sea birds glide overhead on thermals. The dead calm water surrounding our vessel is a deep blue, the color not of shallow water but of deep seas where there are few surface fishes for divers to see and photograph.

But I know this scene is deceiving. The rocky plateau of Ras Muhammad, toward which the sun is dropping, marks what many scientists and scuba divers consider to be the best dive site in the Red Sea, or perhaps even the world.

I think back to Dr. Eugenie Clark's eloquent presentation on Ras Muhammad at CEDAM's meeting last year to select the Seven Underwater Wonders. "If I could only dive in one place in the world," she said, "I would choose Ras Muhammad." Dr. Clark is a professor of zoology at the University of Maryland. Her words were backed by four decades of underwater research in the Red Sea, much of it at Ras Muhammad, where she has documented the un-usual creatures that inhabit a sandy underwater plain below the reef. In her CEDAM presentation, Dr. Clark described one of the richest and most diverse

Like Lake Baikal, the Red Sea sits astride a rift in the earth's crust. Formed over millions of years as the Ara-bian and African continental plates drifted apart, the Red Sea is one of the oldest segments of a huge rift system that stretches from Israel to Mozambique. Using ROVs and manned submersibles, researchers have discovered geo-thermal vents and patches of the basaltic material of oceanic crust–evidence that the sea will continue to widen.

Facing *Hard corals, the main reef builders, construct the foundations upon which soft corals, sponges, algae, and other stationary inhabitants of the reef anchor themselves. On Ras Muhammad's steep underwater slopes, abundant sunlight stimulates these species to crowd available spaces, creating marine panoramas of stunning beauty.*

marine environments in the world—a multicolored coral wonderland, home for thousands of species.

Gorgeous as they are, the reefs of Ras Muhammad represent only a fraction of the coral riches in the seas of this desert region. In the Red Sea, exceptional reef communities are found from the Gulfs of Suez and Aqaba to the Gulf of Aden beyond the sea's southern end. In fact, at the Seven Wonders meeting, Ernie Ernst of the New York Aquarium nominated just the southern Red Sea, while former astronaut Scott Carpenter nominated the whole sea and part of the Gulf of Aden. For her part, Dr. Clark narrowed her choice to the single site she believes best represents the area's marine splendors: Ras Muhammad, whose harsh coast we are now inspecting. In the end, we compromised by including all of the northern Red Sea.

During that meeting, I learned what makes the Red Sea so hospitable to corals and so rich in other marine species. One reason is sunshine; the Red Sea lies in the middle of one of the world's sunniest, driest, hottest places. So scarce are clouds here, says Scott Carpenter, that astronauts orbiting earth in the NASA space shuttle can almost always see the triangular shape of the Sinai Peninsula and even the little outstretched hand at the end that is Ras Muhammad.

At sea level, that means an overabundance of solar energy for corals and their light-loving, photosynthesizing zooxanthellae guests—more than reaches many tropical islands, where thundershowers and clouds reduce light levels for several hours each day.

With all that sunshine come the consistently warm water temperatures on which corals rely. Although nighttime winter temperatures in the deserts of nearby Egypt and Saudi Arabia may dip below freezing, the landlocked Red Sea remains very warm throughout the year; in fact, with average year-round surface temperatures of over 80° F in many places, and average deep-water temperatures (below sixteen hundred feet) of over 70° F, it is one of the hottest corners of the world's oceans.

The Red Sea's geological history also contributes to its variety of species. At one time, the Red Sea opened onto the Mediterranean at its north end, and Mediterranean species flourished here. Then, as the African and Arabian tectonic

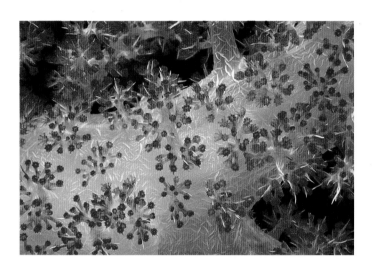

Alcyonarian soft corals usually grow under coral overhangs, but some species prefer bright sunlight. Unlike hard corals, soft corals do not have a hard external limestone skeleton. Rather, they have calcareous white spines, known as spicules, which strengthen the colony's structure. Soft corals grow by adding new branches or by extending existing ones. More than two hundred species have been documented in the Red Sea.

Facing *Despite their delicate appearance, sea fans thrive in strong currents where plankton is abundant.*

plates shifted, the northern straits closed, new straits opened at the southern end, and the Red Sea became an offshoot of the Indo-Pacific, the ocean system where the most coral species are found. Thus, although the Red Sea is dominated by species from the Indo-Pacific, a few Mediterranean species are still found here.

Several million years ago, the Red Sea opened into the Mediterranean Sea rather than the Indian Ocean, and Mediterranean fishes and corals lived in its waters. Subsequent tectonic shifts reversed the openings, bringing in species from the Indo-Pacific, the most biologically diverse of the world's oceans.

ANCHORED SO CLOSE TO ONE OF THE WORLD'S GREATEST REEFS, I'M more than a little anxious to go scuba diving, but jet lag and a hot three-hour taxi ride through the Sinai Desert from Elat to Sharm al-Sheikh have left me too tired to put on a full wet suit, weight belt, and heavy tank. I decide, instead, to go for a quick snorkel before the sun sets.

The coral reefs I've dived on in the Caribbean, Atlantic, and Pacific usually stretch out in all directions as far as the eye can see. This reef is quite different: a single plateau, somewhat rounded, about fifty feet in diameter, rising to within a few feet of the surface. Without my scuba-diving gear, I feel a tremendous sense of freedom as I float effortlessly over a lush hard coral garden on the reef crest.

Parrotfishes, butterflyfishes, wrasses, and scores of smaller fish are also exploring the reef below. Some search for a meal, while others investigate nooks and crannies as possible lairs in which to spend the night, safe from nocturnal predators.

If I had to guess, I'd say there are about three hundred to four hundred fish here from some forty or fifty different species—a greater concentration of marine life than I've ever seen on any Caribbean dive. Without my trusty waterproof fish identification guide, which depicts many Red Sea fishes in color, it's hard to distinguish the subtle differences between species, especially the parrotfishes. I make mental notes of the fishes' colorations and patterns, which will help me identify them when I'm back on board the *Colona II*.

I suddenly realize I'm not alone at the surface. One- to two-foot-long needlefish are all around me. Their blue-green backs and silvery white undersides make it hard to distinguish these sleek animals from shadows just underneath the surface—which is why they hang out there. I see several fish, and then they are gone, blending in perfectly with the wave patterns. Then they come back into view, only to disappear again. As always, I'm amazed at the camouflage techniques developed by reef creatures for protection from their natural predators.

Surgeonfishes and smaller rainbow wrasses cruise the shallows of a Red Sea reef. Here, just a few feet below the surface, divers will also encounter several hard coral species.

Facing *Blue-cheek butterflyfish usually live in pairs, as do many species of reef fishes. Feeding on a diet of coral polyps, blue-cheeks rarely stray far from their home territory.*

After the sun sets, many of the reef's nocturnal invertebrates leave their daytime retreats to feed. Soft corals seem to be a favorite gathering place for crabs, shrimps, and starfishes. Some of these animals are easily detected, like the tan coral crab and tan brittle starfish, **top**. *Others, such as the fingernail-sized red coral shrimps,* **above**, *are almost invisible.*

Facing *So many fishes inhabit some parts of the reef that diurnal and nocturnal species sleep in the same hideaways.*

Perhaps the most plentiful fishes here are the butterflyfishes. Fourteen of these intricately colored species live in the Red Sea; seven are endemic. Most butterflyfishes mate for life and patrol distinct territories, swimming in and out of the reef as they graze on coral polyps.

I try to get close to the parrotfishes and the angelfishes, but they are shy, racing to safety within reef crannies when I get within a distance of about three feet. One rainbow wrasse seems quite inquisitive, swimming up to my face mask, checking me out, then darting away. It comes back quickly, this time with a friend. The two fish accompany me for a few minutes, and then disappear among the myriad of other reef dwellers.

Soon the scene will change as diurnal fishes retire into the reef, making way for the nighttime predators. It's time for me to get back to the boat and ready my camera and scuba gear for tomorrow's dive.

.   .   .   .

IT'S ANOTHER CLOUDLESS DAY ON THE RED SEA, CALM, WINDLESS, AND peaceful. The morning sun is just beginning to peek over the desert mountains, welcoming us to our first dive here.

My objective today is simple but challenging: to find one image which illustrates the essence of the Red Sea, an image like the vast soft coral gardens and spur-and-groove formations of Belize, the sea lions frolicking in the shallows near the Galápagos Islands, or the sponge-and-algae meadows of Lake Baikal.

The site Captain Freddy recommends for my picture is Sting Ray Station at Alternatives Reef. Freddy, Susan, and I descend like a trio of parachutists—spread-eagle—to a depth of forty feet. The visibility is not great, maybe fifty feet, and the site looks pretty barren. Only a few small coral heads dot the sandy bottom. I don't see many fish. I look inquisitively at Freddy. He has obviously seen this expression in a diver's mask before, and signals us to follow him.

We swim for three minutes against a slight current, and then a massive coral column slowly comes into view. It's about twenty feet in diameter and about sixty feet high, rising from the sand to just below the surface.

The massive coral columns at Sting Ray Station, Alternatives Reef, are composed of layer upon layer of dead coral, deposited over hundreds of years.

Facing *Schools of glassy sweepers hover under coral ledges in the Red Sea. This school was found on the coral column above.*

As we move closer, I see hundreds and hundreds of anthias, also called lyretail coralfish. They are constantly in motion, feeding on the zooplankton that drifts by. I have never seen so many fish in one place before.

Through the dense school, we can see soft coral colonies—red, purple, and white alcyonarians—that cover the reef like a Persian carpet. Hard corals, including fire coral and star coral, are here too. This is the scene I'm looking for. I shoot an entire roll of film, hoping to capture my feeling of awe.

Swimming clockwise around the column, I come upon a table coral three feet in diameter. In its shade, hundreds of gold sweepers are swimming in a tight school. When I take a picture with my second camera, the flash momentarily frightens the fish. In unison, they dart away, creating an underwater implosion. After a few more flashes, the fish become accustomed to the burst of bright light, and go back to normal, hovering in perfect formation.

We make several dives at Alternatives, and each visit reveals new treasures: two lionfish swimming upside-down, their white, venomous dorsal spines angled toward us, warning us to stay away; a small moray eel, who looks vicious, but is actually quite docile; and on the reef crest, five feet below the surface, a good representation of Red Sea fishes, perhaps forty to fifty species.

As we swim back to the boat along the sandy bottom, dotted with small coral heads, I'm suddenly attacked by one of the most aggressive fish in the sea. The creature darts out from its lair and charges repeatedly at the lens of my camera. Since this action doesn't drive me away, the fish begins attacking my mask. I blink as it taps briskly at my faceplate.

Fortunately for me, my attacker is only two inches long. But its small size does not deter the two-bar anemonefish, a member of the damselfish family, from trying to drive away intruders, large and small alike, from its territory.

When I swim toward the two-bar anemonefish for a photograph, it retreats into the water-filled tentacles of its partner in life, the Red Sea anemone. The anemone's tentacles, which sting and capture many small fishes, have no effect on the anemonefish. No one is certain how the anemone benefits from the anemonefish's presence; for its part, the little fish is well protected from predation by its host's virulent tentacles.

SUSAN AND I LIKE TO DIVE WHEN THE WATER IS CALM. A TWO-to-three-foot reef report is good news: getting in and out of the water is easy, and more light penetrates the surface, making it brighter for photography down below.

Outside the reef today, there are four- to five-foot seas with occasional seven-foot swells. It's not going to be an easy entry. With Freddy and first mate Steven, Susan and I pile into the inflatable Zodiac and leave the calm waters behind the reef where the *Colona II* is moored.

Going against the wind, we pass through a natural cut in the reef and immediately begin smacking into high waves. We're getting bounced around, and so are my underwater cameras. It's not a good feeling, but it doesn't last long. Within five minutes, we're at the dive site, the Cut at Shaab Mahmud Reef.

Despite surface conditions, the water is surprisingly clear sixty feet down: visibility is about a hundred feet. As I get my bearings, I notice that the reef—lined from top to bottom with colorful hard and soft coral formations—is passing by rather quickly. We are in a strong current that runs along the outside of the reef. Exerting no effort at all, we begin our scenic reef tour.

We pass fish and they pass us. A school of maybe twenty jacks swims by effortlessly in the opposite direction, against the current. Then we are hurled through a school of a dozen or so three-foot-long barracudas. They certainly look menacing, with their protruding lower jaws exposing razor-sharp teeth. In reality, barracudas couldn't care less about attacking divers. We simply are not part of their instinctive diet. But barracudas' instincts do tell them to attack when threatened, and divers should always be cautious.

As we approach the cut, the current picks up speed. Suddenly, we are whisked around the opening like passengers on "The Whip" at an amusement park. We are in shallow water now, not deeper than fifteen feet. A large sand patch lies beneath us, and the current has subsided.

As I kneel on the sandy bottom, I spot dozens of gobies poking their heads out of the sand. More than fifty goby species have been found in the Red Sea, making it hard to distinguish the subtle differences between them. As I look closer at one of these two- to three-inch-long fish, I notice a tiny snapping shrimp inhabiting the same burrow. This is yet another example of a symbiotic relationship: the fish's superior eyesight detects predators from a distance, while the shrimp keeps the burrow clear of sand and sediment. It's a good match, one that helps both animals survive in this world of constant predation.

I swim very slowly over the sandy bottom, taking time to look for other inhabitants of this area. These deceptively barren-looking sand patches conceal some of Ras Muhammad's most fascinating inhabitants. For twenty years, Dr. Clark has made the sand slopes below the reefs the focus of her studies, and she writes about them with rare insight:

> On the open, exposed sandy bottom of the Red Sea, fishes must use more tricks to hide from enemies. Exploring that seemingly barren expanse is like playing the children's game of finding hidden faces in a picture; the longer you look, the more faces you discover.
>
> On one sand slope, ten thousand garden eels live in burrows lined with mucus. To feed, they stretch their willowy, silvery grey bodies almost full-length out of the burrows. Arching their bodies, eyes alert for planktonic food, they curtsy and dip, bow and sway, performing a dance to the orchestration of invisible currents, like cobras spellbound by a snake charmer's flute.
>
> Sometimes a tiny polka-dotted snorkel sticking up from the sand is your first clue to the presence of the shark-repelling Moses sole. Or a tiny opening in the sand lined with four canine teeth turns out to be the breathing hole of one of several species of Houdini-like razorfish— now you see them, now you don't.
>
> If you look closely at the sand where garden eels have disappeared, you might notice that grains of sand are jitterbugging crazily like the prelude to some miniature volcanic eruption. You would have to put these dancing sand grains under a magnifying lens to see the transparent "gypsy" amphipods that hop about carrying domiciles they

Facing *Moray eels have razor-sharp teeth and a firm grip. During the day, they usually hide in tiny reef caves. Although moray eels are far less aggressive than their looks suggest, they will attack divers who provoke them.*

build from sand. My student, Carol Falck, found domiciles with separate "rooms" for a mother and up to ten of her babies.

Among the scores of species discovered in the Red Sea, Dr. Clark herself has found over a dozen. She named one evasive five-inch-long sand-dweller *Trichonotos nikii* for her youngest child, Niki, who was with her when she first encountered the new species.

"Tricky Niki" lives in large colonies. "Sometimes," she writes, "you look at 'barren' sand and realize that hundreds of miniature noses are protruding and twice as many eyes are watching you." During the day, swarms of hundreds or even thousands of tricky nikis swim up into the water column and feed on plankton. "At the slightest hint of danger," says Dr. Clark, "they disappear into the sand like tiny bolts of lightning." At day's end, each male returns to his territory on the sand slope, where he gathers his harem of females and they withdraw into the sand.

At Ras Muhammad, *T. nikii* lives in harmony with the garden eels, sharing the same territory and feeding on plankton. "Their table manners are impeccable," says Dr. Clark; "they never bump into each other or grab for the same piece of plankton." Their territories often include the burrow openings of one or two eels.

About a dozen species of scorpionfish live in the Red Sea. With its algaelike appendages and coloring, this predator blends in with the background, waiting motionlessly for smaller fishes to approach. When one does, the scorpionfish opens its mouth and lunges forward, swallowing its prey so quickly that divers rarely see the moment of capture.

Dr. Clark and her students have gotten to know the inhabitants of the sand slopes so well that they have names for many of them. "Gill," for example, was the dominant female razorfish in a harem of four. "Specimens like Gill are rare," says Dr. Clark. "My diving buddy, David Shen, could count on Gill coming from the deep to greet him even after a year's absence."

Because dominant female razorfish sometimes change sex and become male, Dr. Clark and her researchers were interested in examining Gill's sexual organs to see whether signs of an approaching change were apparent. But when it came time to capture her, David Shen couldn't bring himself to do it. It seemed wrong that scientific curiosity should result in the death of a subject. "We have drawn the line now on all the fish we study," reports Dr. Clark.

Unless it's absolutely necessary to collect and dissect a fish to under-
stand what we are doing, we let nature take its course and decide when
the lifetime of a fish is finished—a far cry from my youthful attitude when
I wrote my first book, Lady with a Spear.

Instead of collecting the fish, we collect their photographs. For our
study of a group of male sandperch, who have lines on the sides of their
faces as distinctive as fingerprints, we've made left and right mug shots of
each male's head on a small plastic card. Year after year, we use these
cards to identify individuals and learn about changes in their territories,
the size of their harems, or their possible replacement or demise.

As I drift through the quiet waters of the sand slope, I feel admiration for Dr. Clark, her researchers, and the world they have discovered beneath the sand. There is not a lot of action or color here—the things a photographer usually looks for—but that doesn't mean it lacks the ingredients to inspire a sense of wonder.

I encounter a three-foot-long crocodilefish, a flat, elongated, bottom-dweller that does indeed resemble a crocodile. Unlike its namesake on land, this animal is nothing to be afraid of; it is shy and rather unapproachable. Its sand-colored, pebble-patterned body helps conceal it from larger predators. If I hadn't seen it move from one spot to another, I would have missed it. I approach the animal slowly, managing to get a single picture before it scurries away into the distance.

Swimming toward even shallower water, I encounter lizardfish, manta rays, sole, and tiny crabs—all camouflaged to blend in with the surrounding environment. Like the crocodile fish, they betray their presence only when they move.

When we surface, we're behind the reef again and the seas are flat. Freddy has expertly guided us back to within twenty yards of his ship, so we have a short, easy surface swim.

The feather starfish, a nighttime feeder, traps floating plankton in its plumelike arms while perched on a coral pinnacle. Once caught, the plankton are transferred by cilia along grooves in the arms toward a central oral opening.

· · · · · · ·

TWO HUNDRED THOUSAND YEARS AGO, THE PLATEAU CALLED RAS Muhammad (the "head"—or cape—of Muhammad) was a submerged coral reef, teeming with marine life. Today, this fossil reef rises several hundred feet above the water, a stark headland of sand and rock.

Just as the branches of a tree may harbor birds, frogs, and insects, the branches of alcyonarians are home to gobies, shrimps, and other small marine species.

Facing *Humphead wrasses, also known as Napoleon wrasses, cruise the coral walls of Ras Muhammad. This five-foot-long individual, known as George, frequently approaches divers, hoping for a hard-boiled egg or other treat. Normally, Napoleon wrasses eat mollusks and crabs, which they crack open with their powerful jaws.*

At the foot of the headland lie some of the most spectacular reefs in the Ras Muhammad area—and therefore, in the world. We will dive today at an offshore underwater plateau almost identical in shape to the ancient plateau that forms the cape. Here the reef crest breaks the surface at low tide, uncovering brown, tan, and yellow soft corals that can survive brief exposure to air.

Our dive team slowly descends to a depth of thirty feet. Freddy signals Susan and me to follow him. Gradually, the massive wall comes into view. It stretches left, right, and down for as far as I can see—a hundred feet in all directions. As we approach the wall, we begin to see that colorful hard and soft corals take up every square inch of space.

The wall is teeming with fish; virtually every fish in the Red Sea can be found here. I feel like I am swimming in a massive aquarium.

The fish seem less timid here than the fish we have seen in other parts of the Red Sea, perhaps because they are used to the divers who frequent this area from the Sharm al-Sheikh day boats.

I hear Susan calling me underwater—an ability I've developed over eleven years of diving with her. I turn, and see that a five-foot-long humphead wrasse is circling her. This great green fish, known as George to Red Sea divers, isn't aggressive. He's only looking for a handout.

George stays with Susan for several minutes, examining her with silver-dollar-sized eyes that move in all directions like the turret on an army tank. When this two-hundred-pound fish realizes that he's not going to get a treat, he moves on to find a more cooperative diver.

Amusing as George's behavior is, it points to the problems posed by human visitors to Ras Muhammad. Feeding fishes may seem like a harmless amusement. But when we treat fish to boiled eggs, cheese dip, and other human fare, we overstep the line between ourselves and these wild creatures. That carelessness can lead to far more damaging infringements on the reef's well-being.

Ras Muhammad's beauty is also its weakness, for the multitudes of divers it attracts place a severe stress on the reef. Not all dive boat operators are as conscientious as Captain Freddy; a carelessly tossed anchor can, in seconds,

Susan Sammon explores a coral canyon at the Temple, a dive site in the Gulf of Aqaba.

break apart huge coral formations below, formations that have taken dozens or hundreds of years to grow. Divers, too, can damage the reef, scraping against it or breaking off a delicate new growth with a swipe of a fin. Most dive boats use permanent moorings around the cape, and many dive guides counsel their clients in responsible diving. But the sheer volume of visitors guarantees that accidents will happen, and with each accident, another piece of the reef ecosystem is lost—for years, decades, or centuries.

Tourism is not the only threat to the future of Ras Muhammad. Thousands of feet below the surface, at the geothermal vents and hot brine pools where the Red Sea's geological future is taking shape, vast quantities of precious metals are distributed throughout the muddy sediment that precipitates from the vents. According to Dr. Clark, in some of these hot spots, "there are five thousand times more iron, twenty-five thousand times more manganese, and thirty thousand times more lead than in normal sea water." The Soviet scientists who dove to six thousand six hundred feet estimated that in the upper thirty feet of the sediment on which they halted their descent, there was over two billion dollars' worth of silver, copper, and zinc alone.

In 1979, engineers with Preussag, a German mining company, succeeded in pumping up some of this metal-rich mud. They then teamed up with the governments of Saudi Arabia and the Sudan to mine the minerals commercially.

There is one difficulty with the venture: the mud in question is loaded with extremely toxic heavy metals in which no life forms more advanced than bacteria can live. For fifteen thousand years, these poisonous sediments have lain undisturbed a mile below the sea surface, never mixing with the waters above them.

Preussag proposes to pump up one hundred thousand metric tons of mud each day to a mining vessel, where the minerals would be concentrated into 1 percent of the mud's total volume. The other 99 percent would go back in the water.

Dr. Clark tells me that these tailings, containing plenty of toxic heavy metals, pose a major environmental risk. Preussag proposes to release the tailings far below the surface. They have conducted studies on plankton populations and believe that the toxic tailings will be adequately diluted. Despite their assurances, it is difficult to feel at ease with the volume of mud they propose to handle. Says Dr. Clark, "I worry where these pilot studies of the 1980s will lead."

Not all the news from Ras Muhammad is this worrisome. As of this year, visitors will find themselves at a very well-run national park—one of a tiny handful of underwater parks worldwide.

Eugenie Clark can take some credit for the existence of this park, for she helped bring the treasured reefs to the attention of Egypt's president in 1980, shortly before Israel completed the return of the Sinai Peninsula (including Ras Muhammad) to Egypt.

The Israeli government had taken good care of Ras Muhammad, controlling commercial and sport fishing in the area. Because the Sinai region was under military occupation, it was not easy for tourist/divers to get there. Once they did, accommodations were Spartan. In the eyes of reef connoisseurs, however, Ras Muhammad's remoteness only added to its allure.

Dr. Clark and many other friends of the reef hoped to ensure that Ras Muhammad would remain protected from harmful fishing practices after its return to Egypt. The reefs also needed a major cleanup. Since the reopening of the Suez Canal, they had become strewn with garbage from passing ships. Once already, Dr. Clark and Ayman Taher, a noted Egyptian underwater photographer, had organized a cleanup day at Ras Muhammad. With the help of a dozen young divers, they hauled up anchors, chains, ropes, tin cans, and carefully unwound and extricated hundreds of yards of monofilament lines and hooks from the delicate corals.

The trash filled up a truck. "But the cleanup of one day was barely noticeable," recalls Dr. Clark. "We were discouraged by the immensity of what was still to be done."

Next, Dr. Clark got help from a diver named Gamal Sadat, who happened to be the son of Egyptian president Anwar Sadat. Gamal showed his father a *National Geographic* article about Ras Muhammad by Dr. Clark and photographer David Doubilet, and then arranged a meeting between the president and Eugenie Clark.

Abundant year-round sunshine helps to foster the tremendous variety of marine life in the Red Sea. Rain storms are extremely rare, but from time to time, dust storms from nearby deserts block the sun for a few hours and turn the sky a dull grey.

Sadat responded enthusiastically to Dr. Clark's appeal. Right then and there, he promised to issue a decree establishing a national park at Ras Muhammad. A week after the meeting, Dr. Clark crossed the Sinai in a convoy of Egyptian government cars full of scientists, lawyers, and bureaucrats who were to confer immediately with the Israeli officials in charge of Ras Muhammad and make plans for the new park.

Within a year, however, Sadat had been assassinated. Dr. Clark soon learned that Sadat's promise of a presidential decree was not worth a great deal among Egyptian politicians. For the park to become a reality, it would have to be approved by the Egyptian Parliament.

By now, conservation groups in Egypt, in England, and in the United States had joined the drive to save the great reefs. Moreover, before his death, Sadat had enlisted support for the project from an influential friend in Parliament, Sayed Marei. On July 20, 1983, the Egyptian Parliament declared Ras Muhammad the nation's first national park.

Even so, the future of the reefs remained uncertain. "The next few years were disappointing," says Dr. Clark.

> Funding for the park was meager. There was no real way to enforce the laws to protect Ras Muhammad. The first park director stayed in Cairo and never understood what was going on at Ras Muhammad. The resident manager stayed in Sharm al-Sheikh and only took the park's jeep out to look at bird life. Spear fishermen from Europe had a heyday. Shell collectors and souvenir hunters started stripping the reefs close to shore.

This depressing state of affairs continued until 1989, when a new resident manager, a retired young brigadier general named Omar Hassan, began to make his influence felt. Working closely with Dr. Michael Pearson, project director for the Ras Muhammad Marine Park, a three-year project being supported by the European Economic Community, he made plans for a visitor information center, clearly marked access points for divers, systematic patrolling of the reefs, and many other improvements. The original park was also enlarged to include two hundred square miles—seventy on land and one hundred twenty underwater, including coral islands in the Strait of Tiran.

The body of this coral crab is no larger than the eraser on a pencil. Hidden within the reef during the day, the tiny creature scavenges for its meals at night among the branches of soft corals.

OUR TIME AT RAS MUHAMMAD IS RUNNING SHORT. IF EUGENIE CLARK has been diving here for four decades and still prefers it to any other site in the world, how could we begin to get to know its splendors in a week-long trip? With every day I spend here, I become more grateful for what she and many other friends of Ras Muhammad have done to guarantee future divers the chance to enjoy this place.

Captain Freddy has two more dive sites for us to visit: "The Temple," known for its plentiful nocturnal inhabitants, and the Shark Observatory, not far from the sand slopes studied by Dr. Clark.

As we don our gear for our night dive at the Temple, Captain Freddy tells us, "Don't be surprised if you run out of film before you run low on air."

He's right. Immediately upon entering the water, the beams of our underwater lights illuminate a colorful reef crest teeming with dozens of nocturnal animals. We are floating over one of the two natural spires that have inspired this site's name.

Divers commonly see schools of fifty or more jacks cruising near Red Sea reefs. Swimming beneath the school reveals the jacks' sometimes circular swimming patterns.

Near the surface, our lights shine on a large basket starfish, its seven thousand arms spread out like a huge Chinese fan. As we approach, we see a tiny silverside struggling to escape the animal's clutches. But it's too late. Strengthening its grip on its prey, the basket starfish draws it closer to its mouth. Soon, the silverside is completely enveloped. As we back away, we see two more silversides caught in the animal's clutches.

In the distance, I see tiny flashes of light, first in one spot, and then another. I move to the area and take a close look, but find nothing. What was it?

Later I learn that what I saw were two flashlightfish. An elliptical, luminescent organ filled with bacteria generates the light beneath their eyes, going on and off when a flap of skin that covers the organ is raised and lowered. Scientists say the light attracts zooplankton prey and helps the fish see the prey in the dark. In addition, flashlightfish may use the light to communicate with other flashlightfish and to confuse predators.

As we move on, we encounter sleeping masked pufferfish, Moorish idols, soldierfish, parrotfish, and several different species of crabs. This is a good time for fish portraits, when the fish are asleep and unaware of our presence. Just as Freddy predicted, I'm out of film long before I've taken all the pictures I want.

·   ·   ·   ·   ·   ·   ·

FOR OUR LAST DIVE, CAPTAIN FREDDY WANTS US TO SEE SOMETHING "really different." And indeed, the wall under the Shark Observatory turns out to be very different from any site we've visited.

Our Zodiac drops us about ten yards from shore. After clearing my mask, I see that we are surrounded by a school of perhaps three hundred lunar fusiliers, slender, four-inch-long blue fish. At first they swim by at a leisurely pace. Then a school of eight jacks moves in, trying to single out weaker fish for a meal. One of the jacks makes its move, and the school suddenly changes direction. Another jack strikes, and the school, in tight unison, changes directions again. We are completely surrounded by the lunar fusiliers and jacks. It's quite exciting to see the hunt at such close range.

As we swim near shore, we find ourselves under a coral ledge. It's a dramatic scene. Sharply defined shafts of sunlight are streaking through holes in the ledge, and schools of fish swim in and out of darkness as they graze on the reef.

Exploring the wall, we find dozens of small caves, lined with red, green, and yellow algae. There are no coral colonies in the depths of these caves, for without sunlight, these animals cannot survive.

In one cave, we are startled to find ten lionfish resting on the walls, ceiling, and floor. I move in to take a picture of these well-camouflaged animals, and they slowly begin to move away. My flash fires, I back off, and they go back to their resting place.

I can't say exactly why, but this is the best dive of the trip. It doesn't matter that I only see one white-tip shark here, not the groups I've been expecting at a place called Shark Observatory. Maybe it's swimming under the overhanging reef, or exploring the caves. Or maybe it's the nitrogen that has built up in my blood stream, inducing a slight touch of nitrogen narcosis—a feeling of euphoria. Whatever the reason, it's a wonderful experience.

Exploring caves and overhangs in the Red Sea, divers must watch for dangerous marine animals such as lionfishes, whose venomous dorsal spines inflict intense pain.

Many Red Sea caves have thin limestone roofs, the remains of extinct coral reefs. Penetrating these roofs, sunshine reaches corals on the sea floor even at a depth of seventy feet.

During the past six days, I've made thirty dives. I'm almost halfway through the photo documentation phase of the project, having dived and photographed in Belize, Baikal, and now the Red Sea. Only four sites remain: Australia, Palau, Galápagos, and the deep ocean vents. I'm looking forward to those trips, and to seeing the images from Emory's and Dan's deep ocean expeditions. Before then, I'll spend many hours going through thousands of slides and making sometimes tedious preparations for the next expeditions. I don't mind, though. The Seven Wonders project has become a labor of love.

POSTSCRIPT: While the book is in production, I learn from Eugenie Clark that Ras Muhammad National Park has continued to prosper. She writes, "Our dreams to protect this magnificent area and yet share it for all to see and enjoy seem to be coming true." She describes a visit in the summer of 1991, when the directors, Omar Hassan and Michael Pearson, proudly took her for another tour of Ras Muhammad. Seven rangers (soon to be increased to fifteen), all trained scuba divers, patrol the park. A visitors' center which blends in with the desert landscape is complete, with audiovisual and exhibition rooms, a library, offices, facilities for the handicapped, and a restaurant overlooking a reef.

The cleanup which began in 1989 has continued. The beaches are pristine, and the Mangrove Channel, where reef fishes spawn and birds rest on migrations, has been cleared of debris and oil from Suez traffic. Most important, hundred of mines laid down during times of hostility between Egypt and Israel have been cleared, opening parts of the reef which have been fenced off for years.

"Every month the rangers sweep the park clean," Michael Pearson told Dr. Clark. "When people look and say, 'What have you done here,' we say, 'If you don't see it, then we have been successful.'"

Entering an underwater cave in the Red Sea, divers often swim through clouds of tiny reef fishes—fairy basslets, reef goldfish, and glassy sweepers.

Facing *Schools of reef goldfish hovering over coral heads in the Red Sea can number in the hundreds. When a predator approaches, the school darts into the reef at lightning speed, only to return to its hovering position once the danger has passed.*

THE GALÁPAGOS ARCHIPELAGO

Nothing could be less inviting than the first appearance of the islands. A broken field of black basaltic lava, thrown into the most rugged waves, and crossed by great fissures, is everywhere covered by stunted, sunburnt brushwood, which shows little signs of life. The dry and parched surface, being heated by the noon-day sun, gave to the air a close and sultry feeling, like that from a stove: we fancied even that the bushes smelt unpleasant...

I'M READING THE ENTRY FOR SEPTEMBER 17, 1835, IN THE JOURNALS OF Charles Darwin, one of the most celebrated naturalists of all time. His first impressions of Chatham Island (now San Cristóbal) in the Galápagos Archipelago sound pretty disagreeable. But the discomfort of a hot morning on a remote tropical island was nothing unusual for the twenty-six-year-old Englishman, who was in the middle of a five-year expedition on the *Beagle*, collecting and studying animal and plant life around the world.

Darwin spent only thirty-five days in the Galápagos Islands, and in that time, he accomplished his goal of gathering a representative selection of the archipelago's plants and animals. His journals, which vividly describe what it is like to visit these volcanic islands, contain detailed accounts of his first encounters with marine and land iguanas, sea lions, and dozens of bird species. As I read, I'm impressed by how Darwin's scientific curiosity invariably wins out over physical obstacles. On meeting the giant tortoises he writes,

The day was glowing hot, and the scrambling over the rough surface and through the intricate thickets, was very fatiguing; but I was well repaid by the strange Cyclopean scene. As I was walking along I met two large tortoises, each of which must have weighed at least 200 pounds: one was eating a piece of cactus, and as I approached, it stared at me

Red sea stars are among fourteen starfish species in the Galápagos Marine Reserve, a protected area that includes all the waters within a line around the outermost points of the islands as well as a fifteen-mile buffer zone beyond the line.

Facing *The pinnacle and submerged crater on Bartolemé Island are vivid evidence of the volcanic forces that shaped the Galápagos Archipelago. The first islands here appeared more than two million years ago when peaks of submarine volcanoes broke the sea surface six hundred miles off the coast of Ecuador.*

and slowly walked away; the other gave a deep hiss, and drew in its head. These huge reptiles, surrounded by the black lava, the leafless shrubs, and large cacti, seemed to my fancy like some antediluvian animals.

Brief as his sojourn in the Galápagos was, Darwin would spend many years thinking about the specimens he collected there. In time, his reflections upon the islands' fauna would become a part of his famous theory of the evolution of species.

In Darwin's day, students of natural history were taught that species were "stable"—they had remained the same since God had created the world. While in the Galápagos, Darwin had been told by the vice-governor that tortoises from different islands could be told apart by their markings. Darwin also observed subtle variations among mockingbirds from different islands. If species were truly stable, these creatures, which came from islands near each other, should have been very similar. The fact that they weren't prompted him to ask whether members of a species, when separated, could begin to change until they had become different species.

Darwin didn't seize upon the theory of evolution all at once. Nor did he construct it only on the basis of what he found in the Galápagos. Back in England after his long voyage, he would spend a quarter of a century analyzing his collections, reviewing specimens from other naturalists' expeditions, and consulting with taxonomic specialists before he published a full account of the theory in *The Origin of Species*—a work which laid the foundations of the biological sciences as we understand them today and which revolutionized the way humanity understands its place in the history of the planet.

Resting on the lava shore, this month-old Galápagos sea lion shows no fear of humans. Lacking natural predators, sea lions here are easily approached. But as numbers of visitors soar, such one-on-one experiences may become rare.

MY FIRST IMPRESSIONS OF THE GALÁPAGOS ISLANDS, IN AUGUST 1988, are quite different from Darwin's. As we cruise south from the island of Baltra, I am amazed at how beautiful each island is, and how much each differs from the others. Some have rocky, jagged peaks, jutting straight up from the shoreline.

Others are almost totally flat, dotted with only a few cactuses and wildflowers.

On many islands, hollow lava tubes running from volcanic craters to the shore testify to the violent eruptions that shaped these islands six hundred miles off the coast of Ecuador. Like the Hawaiian Islands, the Galápagos Islands are the peaks of gigantic underwater volcanoes, created as a tectonic plate moved over a thin patch, a "hot spot," in the earth's mantle. As the plate drifted slowly southeastward, the archipelago's eastern and southern volcanoes left the hot spot and became extinct, while to the west, over the hot spot, new volcanoes rose

Bartolomé Island's desolate volcanic landscape belies the rich and varied marine life just offshore.

The manta ray, whose winglike pectoral fins may span eight feet, feeds on microscopic plankton and small fishes. A pair of flaplike fins on the manta's head helps guide food into its mouth. Remoras, usually seen in pairs, often hitch rides with mantas—as do some divers, tempted by the fact that these Pacific rays lack the stingray's intimidating tail spine. Mantas, however, dislike having humans on their backs, and will abandon dive sites where it happens, depriving other divers of the thrill of swimming near the largest of all the rays in the sea.

Although the nine-inch-long large-banded blenny is one of the largest blennies in the Galápagos, it is also one of the most timid. When approached, it quickly retreats into crevices between lava rocks.

Facing The Gulf tulip shell inhabits shallow reefs and sand beds from the Gulf of California to Peru, delighting divers with its brilliant orange or red body. When the mollusk senses a threat, it withdraws into its shell, sealing the opening with the leathery bottom pad of its foot. The largest mollusk in the Galápagos, the tulip shell can be as much as a foot in length.

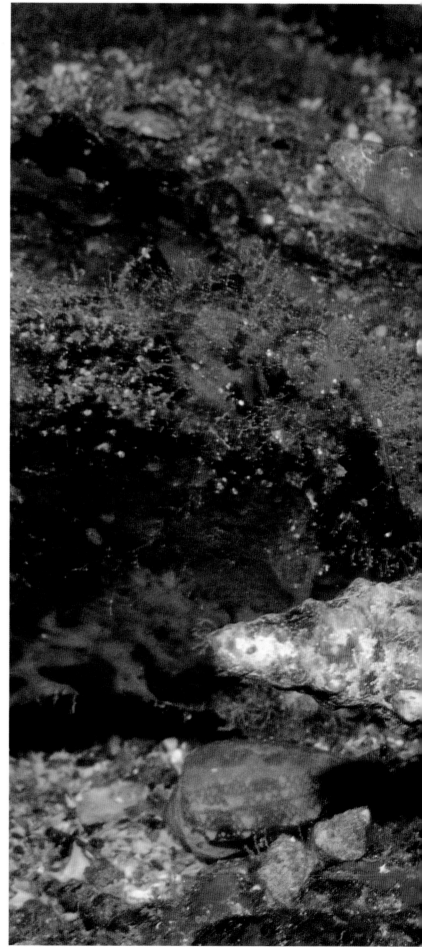

Because blue-footed boobies are near-shore feeders, visitors encounter them more frequently in the Galápagos than the more numerous red-footed and masked boobies. In all, nineteen bird species nest within the archipelago; five are endemic and the rest migratory.

Facing *The Galápagos Islands contain only six mammal species, among which the sea lion predominates. Often confused with their relatives the seals, sea lions differ from them in several important ways. While seals' ears are completely hidden, sea lions' ears include the auricle, the fleshy external flap. They also use their front flippers to propel themselves in the water; seals use their hind flippers. In 1990, the sea lion population in the Galápagos was estimated to be between twenty and fifty thousand.*

from the sea floor to the surface. At the southeast end of the archipelago, the Galápagos' oldest island dates from at least two million years ago, and to the west, the newest islands—still volcanically active—have existed over half a million years.

Today we cruise fairly close to the islands, and can observe sea lions basking in the sun on the sandy beaches and rocky shores. The bulls bark as we go by, warning us to stay away from the newborn pups.

On the cliffs, I see dozens of blue-footed and masked boobies. These sea birds are artful hunters with excellent eyesight. I watch them stare at the clear, shallow, blue-green water below. Suddenly, one takes off. Folding its wings into its body to streamline its descent, it dives into the waves. Ten or twelve seconds later the booby surfaces and takes flight again—with a small fish clenched tightly in its bill.

The Galápagos Islands are home to a wide array of other sea birds, including frigate birds, tropicbirds, petrels, and terns. Five species are endemic: the waved albatross, the Galápagos penguin, the flightless cormorant, and two types of gulls. All are here because the waters of the archipelago contain an abundance of fish.

Nutrient-rich cold currents sustain these fish populations, which in turn sustain the islands' sea birds, sea lions, and fur seals. The archipelago sits astride two cold ocean currents. From the Antarctic, the Peru or Humboldt Current sweeps up the coast of South America, curving westward near the Galápagos, where it becomes the South Equatorial Current. From the mid-Pacific, the Equatorial Undercurrent flows east, underneath the South Equatorial Current. Meeting the flanks of the Galápagos platform, it rises to the surface with its bounty of planktonic life.

From January to March, as these cold currents weaken, the warm, east-flowing North Equatorial Countercurrent shifts southward toward the Galápagos, and from the north, the Panama Current draws warm waters from the Gulf of Panama.

As you can imagine, these currents create turbulent surface and underwater conditions around the islands. They've also brought in an unusual mix of

fishes: tropical species from the Indo-Pacific and Central America, and temperate species from South America. If Darwin could have put on scuba gear and spent time underwater, he would have found some three hundred fish species along the islands' submarine cliffs and slopes.

. . . .

IT'S LATE IN THE AFTERNOON AS WE APPROACH THE FIRST DIVE SITE, a peaceful, shallow bay on Plaza Island. The water is clear; we can see starfish on the sandy bottom twenty feet below.

Our wooden-hulled dive boat, the *Beagle III*, is the former research vessel of the Charles Darwin Research Station, and Susan and I are joined on this expedition by eight other conservation divers. All of us are here for the first time, and the enthusiasm level is high.

Susan and I quickly put on our scuba gear and are the first to enter the blue-green water. It's cool—68° F. We're glad we brought our diving hoods and gloves.

Soon we are greeted by six sea lions—two adults and four pups. They fly toward us from all directions, maneuvering their sleek bodies in playful dives and turns. To our surprise, no matter where they are heading, they keep their eyes fixed on us.

As the sea lions get accustomed to us, they swim closer and closer. We laugh at how they blow bubbles out of their noses as they approach. One pup picks up a pencil sea urchin from the sandy bottom with his mouth and carries it toward us, dropping it just above our heads. Before it reaches the bottom, he snatches it up and bounces it on his nose. A moment later, another pup grips my fin with his mouth, releasing it after he has said hello. These are wild animals, I remind myself. They are only doing what comes naturally.

Back on board, we get a briefing on the Galápagos Marine Resource Reserve from local naturalist María Ramos. María tells us that Ecuador established the reserve with a presidential decree in 1986 to protect the water column, seabed, and marine subsoil of the Galápagos Archipelago.

The marine reserve is the largest of its kind in the Americas, with a total

Because sea lions are not hunted in the Galápagos, underwater explorers encounter them on almost every dive. At Lake Baikal, by contrast, approximately six thousand seals are killed every year for clothing and food, and the species has become virtually unapproachable.

surface area of over twenty seven thousand square miles. Approximately twenty thousand square miles lie within the archipelago's interior waters, defined by a line around the outermost points of the islands; the rest is a buffer zone extending fifteen nautical miles seaward from the interior line.

Marine conservationists in Ecuador and around the world were pleased with the decree. But as I learned from Marsh Sitnik of the Smithsonian Institution, who joined us at our Seven Wonders meeting in August 1989, the decree called for government committees to come up with a management plan. As of 1991, no such plan has been enacted. "Now," writes Sitnik, "growth in three areas—tourism, the local population, and the fishing and lobster industries—poses an ever more serious threat to the marine environment of the Galápagos."

Despite this delay, Ecuador has taken excellent care of the Galápagos Islands and their natural treasures. María tells us about the Galápagos National Park, established in 1959 to protect the archipelago's terrestrial flora and fauna. Ninety-seven percent of the Galápagos' land area is included in the park.

The park benefits from a remarkable partnership between its managers and the scientists of the Charles Darwin Research Station, founded in 1960 at Academy Bay on Santa Cruz Island. What the scientists learn about Galápagos wildlife is quickly shared with park administrators, enabling them to help endangered species and address other problems.

With the help of the research station, the park service also trains its tour-

Schools of blue-striped snappers are common throughout the underwater volcanic ridges of the Galápagos.

Left White-tip reef sharks, like all sharks, are actively and aggressively hunted. In 1990, more than one hundred million sharks were killed worldwide, for sport, for meat, and even for their fins, a sought-after delicacy for soups in the Orient. That same year, fewer than a dozen shark attacks on humans were recorded. Most happened on the surface when the shark mistook the human for a seal pup or a wounded fish. Without protection, some species within this ancient order of fishes—one of the oldest in the oceans—may soon be threatened.

Brown moray eels, **left**, and Panamic green moray eels, **above**, rest in volcanic walls during the day and forage in open water and on sand flats at night. These relatively docile animals look fierce because of their constant open-and-shut mouth movements, a reflex that brings oxygenated seawater over their gills. According to biologist Godfrey Merlen, who has spent years documenting marine life in the Galápagos, approximately sixteen moray species inhabit the archipelago.

ist guides, who accompany every group of visitors to the island when they venture into park areas to see the archipelago's famous creatures. As María tells us, "Being in the Galápagos is like being in a zoo and aquarium . . . The difference here is that you are literally on a one-to-one basis with the animals. 'Look, don't touch' is the basic rule in the Galápagos—on land and under water. You see, the reason you can get so close to the animals is because they are unafraid of people. They are not hunted or harassed, so they have nothing to fear from humans."

Like species on many oceanic islands, the inhabitants of the Galápagos evolved with few or no predators, and their natural defenses—fleeing, fighting, hiding—have diminished at the expense of other capabilities. That was fine until humans found the islands. As domestic pigs, goats, dogs, and cats escaped from settlers and became feral, giant tortoises, marine turtles, land iguanas, marine iguanas, and others suddenly had enemies who effortlessly gobbled up their eggs and young. In some areas, these feral livestock and pets brought populations of native creatures close to zero. In others, humans themselves did the work.

Nineteenth-century whaling ships often stopped in the Galápagos Islands looking for water or provisions. As there is only one freshwater stream in the whole archipelago, their attention turned to the giant tortoises. Potable water could be obtained from their bladders and other body cavities. Their flesh was good to eat. Their fat could be distilled into a fine oil. Most important, the tortoises could live up to a year without food and water, stored in stacks in the ship's hold—a very simple way to keep fresh meat on board.

Whalers took tortoises from the islands by the hundreds. The largest number known to have been stowed aboard a single ship is 350. Over the decades, it is estimated that at least a hundred thousand tortoises were removed from the Galápagos, a deadly harvest that helped reduce the total number of Galápagos tortoises from perhaps a quarter million to a current total of about fifteen thousand. Some subspecies found only on particular islands were wiped out. In the whole archipelago, there were originally fifteen tortoise races. Five

Long-spined sea urchins are animals to avoid. Their sharp spines can puncture a wet suit and embed themselves in skin, where they cause pain for several days. A barb at the end of each spine makes it impossible to remove.

Facing *Whale sharks are the largest fish in the sea, reaching a length of up to sixty feet and a weight of possibly twenty tons. Gentle and slow-moving, these giants are harmless to divers. They feed on plankton and small fishes, which they strain through a mat of long gill-rakers. One sure way to drive away a whale shark is to grab a fin and take a ride. This practice has made whale shark sightings less and less frequent in the Galápagos and throughout the world.*

lived on Isabela Island, and ten other islands each had an endemic subspecies. The tortoises of Santa María and Santa Fe Islands were extinct by 1906. On Fernandina, the California Academy of Sciences removed the last known survivor in 1926. The last known tortoise from Pinta Island now lives at the Charles Darwin Research Station, and two other populations remain dangerously low.

Other creatures of the Galápagos were also hard hit over the years. The fur sea lion population had almost disappeared by 1906, though it has now recovered and numbers over thirty thousand. And on the island of Baltra, you won't find any land iguanas. During World War II, the United States Air Force built an airfield there. A big pastime for servicemen at this uneventful outpost was killing land iguanas. When the United States turned over the base to Ecuador after the war, not one living iguana remained on the island.

I think of my dive among the sea lions and of Charles Darwin's first encounter with giant tortoises. As an underwater photographer, I've often been close to wildlife. But rarely has the contact seemed so trusting as my playful swim this morning. It takes a lot of management and a lot of education to ensure that thousands of tourists can experience that immediacy without affecting the creatures they've come so far to see.

Humankind remains the most troubling factor in the future of the Galápagos. Over seventy-five thousand tourists a year now visit the islands, and who could blame them? It's one of the most memorable places on earth. But as the numbers creep upward, so do the pressures—more hotels, more permanent residents to staff them, more cruise vessels, more boat traffic, more trash, more spills, more of everything. Just recently, permission was granted for cruise vessels carrying up to a thousand passengers to tour the islands—a decision which troubles many Galápagos conservationists. I can only hope that the wisdom which has guided the management of the park so far will persist and that the Galápagos Marine Resource Reserve will benefit from it.

Even snorkelers can enjoy the artful movements of Galápagos sea lions. No matter how quickly they turn, twist, and dive, sea lions never take their eyes off underwater visitors nearby.

OVER THE NEXT FEW DAYS, EXPLORING SMALLER ISLANDS FOUND ONLY on the most detailed maps, I learn why diving in the Galápagos is not for novices. Strong currents surround us on many dives, and at times I'm carried backward despite my efforts to swim forward. On several occasions, the currents dictate drift dives; we start from a dinghy, which follows our bubbles as we move along. When we surface, the dinghy is there to pick us up, thanks to our watchful crew.

One of my favorite dives is at Devil's Crown, so-called because of the extremely treacherous underwater currents around the jagged, crownlike circle of black rocks that break the surface here. We enter the water and begin swimming toward a vertical drop-off. At first, it appears that there is not much life on the black volcanic wall, and compared to tropical coral reefs, there isn't. One reason is that corals are not as productive in cooler water.

There is another reason, however, for the absence of lush coral gardens. In 1982 and 1983, a shift in the flow of Pacific Ocean currents raised water temperatures in the Galápagos Archipelago as much as fifteen or twenty degrees, killing 95 percent of the area's corals. El Niño, as the weather event is known throughout the Americas, also drenched the Galápagos with nine times more rain than usual, and brought sometimes disastrous changes to weather patterns in South and Central America.

In the Galápagos, not only the corals died. The high water temperatures sharply reduced quantities of plankton, and the shock ran through the food chain. Among sea lions, few pups born in 1982 or 1983 lived. The fur sea lions suffered even more; most of the seals born after 1980 died. And between a half and two-thirds of the Galápagos' marine iguanas starved to death. Their normal food, red and green algae, couldn't tolerate the warmer water. A species of brown algae which the reptiles found indigestible flourished instead. They ate it, but received no nourishment.

All of these species rapidly recovered with high birthrates in the following years, and according to Pat Whealan, a scientist at the Charles Darwin Research Station in the Galápagos, the corals are also coming back, but slowly.

Even as they do, El Niño currents will strike again. Minor El Niño events

Orange-cup coral is one of a handful of coral species that inhabit the chilly waters of the Galápagos Islands. Currently, corals are recovering in the archipelago after the devastating 1983 El Niño, which brought torrential rains and unusually warm and nutrient-poor seawater to the Galápagos and other South American waters, killing a high percentage of corals.

A keen eye is required to detect many of the underwater wonders of the Galápagos. This inch-long triplefin blenny, for example, was resting upside down on the roof of a small cave at a depth of ninety feet. Without the aid of artificial light, the fish would have been invisible to a diver.

occur every five or ten years, and big ones such as what happened in 1982-83 may develop once in a century. Some theorize that the Galápagos Islands, with their volatile mix of warm and cold currents, are not suited to being an especially productive place for reef-building corals.

As we swim closer to the wall, we see blue encrusting sponges, black corals with delicate green branches, and small colonies of orange-cup coral. Closer still, the wall comes alive with several species of green, yellow, and red algae, an important part of the food chain. Starfish are everywhere, in shades of red, yellow, and pink. Familiar tropical fishes, including parrotfish, butterflyfish, pufferfish, and sergeant majors, swim up and down the wall in search of a meal. Curious blennies, pinky-sized fish that live in small holes, poke their heads out.

We encounter a school of about a hundred king angelfish and yellow-tail surgeonfish. Completely surrounded, we thrill, once again, to being so close to wild creatures. The school moves on, grazing along the wall as if we weren't there.

. . . .

AS VOLCANIC ISLANDS, THE GALÁPAGOS OFFER SOMETHING UNIQUE for our team of scuba divers—cave diving. However, since this is a fairly dangerous activity, recommended only for specially trained individuals, we decide to dive at the mouth of a small cave, agreeing not to go deep inside.

Once inside the cavern, Susan and I turn on our underwater flashlights. Since there is very little sunlight here, there are no hard corals, which must have light to survive. But there are sea lions. They dart in and out of our flashlight beams, performing a ballet that is simply incredible to behold as they appear and disappear in the narrow shafts of light.

Along the cave wall we find slipper lobsters, clawless crustaceans that look like the tail end of the lobster one usually gets in a restaurant. We see more than enough lobsters here for a feast on our boat, but we know they are protected by local conservation laws and we move on.

As our eyes get accustomed to the underwater terrain, we begin to notice scorpionfish. Although I've seen scorpionfish in the Caribbean, the Red Sea, and

Delicate green branches conceal the remarkable beauty of black coral, which is illegally harvested, polished, and carved into jewelry. Despite the prohibition, black coral jewelry abounds in gift shops in the Galápagos and in Ecuador. By learning about corals, shells, and other rare marine species which shouldn't be harvested and sold, conservation-minded travelers can help end the market in these natural wonders.

The scorpionfish of the Galápagos Archipelago are as ugly as their relatives anywhere in the world, and their venomous dorsal spines are as hazardous to predators–and unwary divers.

on many Indo-Pacific reefs, the Galápagos scorpionfish seem to have better camouflage. Sitting still, they are nearly invisible; their rough scales blend in perfectly with the algae-covered rocks as small fishes swim by unawares—perhaps for the very last time.

Like all scorpionfish, these fish have venom glands on their dorsal spines to keep predators away. Meeting this animal, both prey and predator face mortal danger.

As we leave the cave, we see a small school of hammerhead sharks silhouetted against the clearer water outside. They don't even stop to take a look at us. I'm always surprised at how difficult it is to get close to sharks, and equally surprised at the bad reputation these majestic animals have acquired.

. . . .

SEVERAL THINGS MAKE EARLY-MORNING DIVING ENJOYABLE, INCLUDing dramatic lighting, more fish activity, and the simple fact that it's fun to be in the water before breakfast.

In the Galápagos, we make many morning dives, but at Santa Cruz Island there is a special treat. Each morning, we see three- to four-foot-long marine turtles sleeping headfirst under ledges, firmly wedged so as not to float to the surface, where predators might be.

As we approach our first sleeping turtle, we are struck by its beauty and vulnerability. The sound of our bubbles wakes the turtle. Cautiously he backs out of his resting place. We make eye contact, but only for a moment. With powerful thrusts of all four flippers, he glides away, fading gradually into deeper water.

It saddens us to think that in other parts of the world, these docile animals have been slaughtered for jewelry and meat in such large numbers that they are endangered. In the Galápagos Islands, at least, they are protected.

Santa Cruz offers the best visibility we've had, up to eighty feet. Here we also see five-foot-long moray eels swimming in open water, schools of barracudas, groves of black coral, and—for the first time—sea horses wrapped around the branches of sea fans. These are our final dives in the Galápagos. We make three

In the nutrient-rich waters of the Galápagos, barracudas grow to a length of three feet and travel in schools of hundreds.

Facing *Marine turtles swim gracefully in the open ocean, easily outdistancing even the fastest human. On land, where they lay their eggs, marine turtles move slowly, exhausting almost all their energy to bury their clutch safely. Before humans arrived in the Galápagos, the eggs had a chance of hatching. Today, all that has changed; introduced pigs not only have a taste for turtle eggs, but a knack for finding clutches even beneath two feet of sand.*

Top *Marine iguanas are found only in the Galápagos Islands. Between dives in the archipelago's cool waters, they raise their body temperature by sunning themselves for hours on lava rocks near the shore.*

Bottom *Land iguanas are also common in the Galápagos. More colorful than their marine counterparts, they dine on cactus leaves and flowers.*

Facing *A dense row of teeth makes this horned blenny a fearsome predator to the tiny animals that fit inside its quarter-inch-wide mouth.*

dives a day here, cataloguing as many different fish species as possible.

After the last dive, as we are packing up our gear, there is a strange silence on board. We are all saddened that the journey has almost come to an end. Not only will we miss exploring the underwater and terrestrial wonders, but we will miss the camaraderie that develops on adventures like this.

.   .   .   .

NO TRIP TO THE GALÁPAGOS ISLANDS WOULD BE COMPLETE without seeing Bartolemé's Pinnacle, the highest point in the archipelago. Rising straight out of the sea, this natural landmark is visible for miles.

We leave the *Beagle III* and head for land in our dinghy. As we approach the sandy shore, we notice hundreds of marine iguanas clinging to the rocks. These animals are found nowhere else in the world. With specially adapted mouths, they graze on algae on underwater rocks. And thanks to specially adapted claws, they can hang on even in fairly rough seas.

We hike to the top of a nearby peak, overlooking a bay in the shadow of Bartolemé's Pinnacle. From here, I can see for miles; at least a dozen islands are visible. It's late afternoon, and the sun reflects off the calm water, adding a sparkling effect to the breathtaking scene.

As I look back over our Galápagos adventure from this vantage point, my thoughts turn to how important nearby terrestrial environments are for each of the underwater wonders of the world. I think of Lake Baikal, where forest preserves protect only some shores. Where the forests have been cleared, eroded sediments cloud the water. Near towns and factories, untreated waste pollutes large sections of the lake. Here in the Galápagos, the marine environment remains pristine in part because the land is unspoiled.

As a fiery red sun sets, we discuss the life of Charles Darwin. We all wonder what his impressions of the Galápagos Islands would have been if he could have scuba dived here. We all wish he could have explored the underwater wonders of what the Spanish aptly called *Las Islas Encantadas*—the Enchanted Islands.

THE GREAT BARRIER REEF

DREAMTIME: THE NAME AUSTRALIAN ABORIGINES GIVE TO THE ERA when the heroes of their ancient legends lived, a magical time filled with images and events that stir the imagination.

Today, several miles off the coast where the aborigines once thrived, I am in my own magical dreamtime. Gliding ever so slowly through deep blue water, I am surrounded by shafts of light that twinkle in the stillness, a light show that inspires my imagination.

In this liquid environment, eight hundred times more dense than air, I can fly, like the legendary bird-man of the ancient aborigines. I wonder what those tribesmen would think if they could see me today, a strange, amphibious creature who appears to breathe both air and water.

I stabilize myself above the "bommie," a somewhat cylindrical reef rising sixty feet from the sea floor. Branching and boulder-shaped coral colonies dominate the surface of the reef, and an amazing variety of beautifully colored and exotically shaped fish swim among them. As I look more closely at the scene, I realize that several dozen species of fishes and corals are plainly visible. And yet, bewildering as this profusion is, there are dozens, maybe hundreds of smaller creatures tucked into recesses of this complex ecosystem.

The profusion of conspicuous hard corals distinguishes diving on this part of the Great Barrier Reef from my recent Red Sea experience. On most Red Sea walls and coral heads, the first things you notice are the dozens of delicate soft corals, resembling a landscape of flowering trees and shrubs laid out by a master gardener. Here the stony corals' myriad forms evoke a sculptor's workshop. Their shapes are indicated by their common names: brain coral, staghorn coral, and boulder coral.

Above and facing Each coral colony is founded by a single coral animal. From time to time, corals release clouds of eggs and sperm. Fertilized coral eggs become larvae drifting with other plankton. Now and then one of these floating polyps lands on a hard surface and attaches itself firmly. It then multiplies asexually, putting out new polyps as buds. Each polyp begins to secrete the calcareous material of the skeleton as reproduction continues through budding. Layer on layer, the colony rises in a shape determined partly by genetic factors and partly by depth and current.

White bubble coral owes its name to its nocturnal appearance, when bubble-shaped polyps protrude from its limestone skeleton.

Facing Since their first appearance five hundred million years ago, corals have evolved unique defensive mechanisms. Some produce secretions that kill competing corals nearby; others a sunscreening substance that protects them from solar rays when exposed at low tide. Corals' brilliant colors come from pigments in the living tissues of their polyps. When taken from the water, the polyps die and dry up, leaving only a stark white coral skeleton.

The branching structure of staghorn corals, which grow near the surface where wave action is strong, gives them stability. Corals at deeper levels also grow in adaptive shapes; the broad surfaces of table coral, for example, help the colony absorb as much sunlight as possible.

There are many soft corals here too, mostly under hard coral overhangs and table coral formations. These delicate animals may look harmless as their transparent and often luminescent bodies sway in the gentle current. But like all corals, they live by capturing, stinging, and consuming microscopic animals that drift in the current.

As I examine this multitude of coral species, I recall an interesting fact about the Great Barrier Reef. In this ideal setting for tropical marine life, where sunshine is abundant and the water is relatively shallow, clear, and warm, any individual reef will probably have more coral species than are found in the entire tropical Atlantic. The Great Barrier Reef as a whole contains over four hundred coral species—ten times more than the Atlantic. This diversity presents underwater explorers like myself with the challenge of identifying the different species, many of which look quite similar. The highly productive reef also offers scientists a living laboratory in which to conduct research.

The value of coral reef research is illustrated by the work conducted at AIMS, the Australian Institute of Marine Science. Recently AIMS scientists analyzed the protective coating that hard corals secrete when exposed to air at low tide. This coating, which acts as a sunscreen against the sun's harmful ultraviolet rays, helps these aquatic animals survive in air for several hours. If you like sunbathing, you may someday benefit from these findings; a synthetic sunscreen similar to the coral's is now being jointly developed by AIMS and a major pharmaceutical manufacturer.

. . . .

THE PEACEFUL FEELING I GET FROM THIS REEF SCENE IS MINE ALONE, for each fish, large or small, is on constant alert for predators—and on the lookout for its own next meal.

Tube worms dot the surfaces of some hard corals. These very distant relatives of earthworms are sometimes called Christmas tree worms or peacock worms because of their cone shape and bright coloring—blue, yellow, green, orange, or white. As I approach, they retreat into the safety of their protective tubes, warned of danger by their light- and pressure-sensitive organs.

Settling on the sandy bottom, at a depth of fifty feet, I'm suddenly reminded of "Diver Dan," a children's television program from the 1960s. In one episode, Diver Dan gets his foot caught in a giant clam, three feet in diameter and weighing several hundred pounds. As Dan runs out of air, he struggles to free his foot from this monster. With his trusty dive knife, he finally pries his foot loose and swims to the surface—just in time for a commercial.

The reason for this reminiscence is that I'm surrounded by six of these giants. In reality, it would be almost impossible to become trapped by a giant clam. Like the tube worm, a clam senses when something is approaching and closes its shell. Wedging one's fin inside the clam's mantle would be a neat trick indeed. Moreover, these animals feed on drifting microscopic phytoplankton and couldn't care less about chomping on a diver.

The giant clam, *tridacna gigas*, is the largest bivalve in the world and one of the Great Barrier Reef's most enthralling sights. Upon close examination, the clam's mantle reveals bright colors and intricate patterns, caused in part by the symbiotic algae, zooxanthellae, which live within the mantle's flesh.

The algae are partly responsible for the size of *T. gigas* and its somewhat

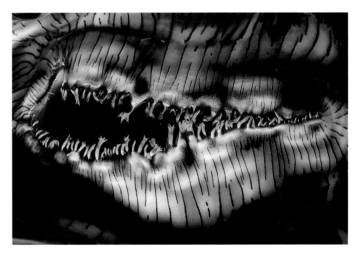

The Great Barrier Reef is filled with many wonders, like the colorful mantle of the giant clam. Specialized algae living within the exposed mantle tissue produce the intricate color patterns. They also produce nutrients for the sedentary giant, which can grow to a length of three feet.

smaller kin, *T. derasa*, *T. squamosa*, *T. croces*, and *Hippopus hippopus*. Like the symbiotic algae in hard corals, these algae manufacture carbohydrates through photosynthesis. The clam absorbs the carbohydrates as nutriment, and produces nitrogenous waste—food for the algae. The cycle is so efficient that the clams can survive for long periods with little input from drifting plankton.

Peering into this sedentary giant, I marvel at the beauty of the animal's delicate siphon, which pumps water in and out of its body. Tridacna clams live a long time—perhaps a hundred years—but appear to grow most rapidly in the first ten years or so. Once they were widespread throughout the shallows of the southwestern Pacific. But in this century they have been heavily overharvested; their adductor muscles are a prized delicacy and their massive shells are a collector's item. Only on the Great Barrier Reef, where they are protected, are they still found in large numbers. Even here Taiwanese fishing fleets illegally strip remote reefs clean of giant clams from time to time.

· · · · · ·

FOR TWO DAYS I'VE BEEN PHOTOGRAPHING THE FLORA AND FAUNA of bommies in the northern section of the Great Barrier Reef. Occasionally called "The World's Largest Living Creature," the Great Barrier Reef is home to some fifteen hundred species of fish and more than four hundred species of coral, so I have plenty to photograph. Stretching for more than twelve hundred miles along Australia's northeast coast, the Great Barrier Reef has also been called "The Largest Barrier Reef in the World."

These designations, however, are technically incorrect. As Dr. Ernie Ernst of the New York Aquarium has told me, the Great Barrier Reef is a "complex of over twenty-seven hundred submerged coral reefs and over eight hundred sandy coral cays and reef-fringed vegetated cays and continental islands."

Ernst describes the complex as including three regions. The northern region, southward from the Torres Strait, includes ribbon reefs, delta-shaped reefs, and "plug" reefs, cylindrical forms like those I saw at the site of my first dive here. "The windward sides of the ribbon reefs in this region," he writes, "plunge into the abyss a mile below. Brushed by strong oceanic currents and pounded by

Above and facing *Pixie Pinnacle, a protected dive site, rises from a depth of about forty feet to about five feet below the surface. Coral pinnacles, known as "bommies" to local divers, dot sandy patches on the Great Barrier Reef, providing a relatively safe habitat for reef fishes.*

Rocky continental islands, the remains of ancient mountains, dot coastal waters at the northern end of the Great Barrier Reef. In this region, fringing reefs form an almost continuous line along the edge of the continental shelf many miles from shore. These are the oldest portions of the Great Barrier Reef, dating back twenty million years. Scientists believe the reefs' central and southern portions are much younger, about a million years old.

trade-wind waves for much of the year, reefs in this area form true barriers. Corals and algae encrustations on the reef fronts are massive and beautiful, and the fish here are large and fast-moving."

Behind the ribbon reefs are fringing reefs around coral cays and continental islands. "Bathed by milder currents and waves," reports Ernst, "the corals, fishes, and invertebrates here are more colorful, delicate, and diverse." Because the area is so remote, it hasn't felt much impact from tourism. Nesting sites for sea turtles and sea birds are common, and the dugong, an endangered sea cow related to the Florida manatee, has one of its last strongholds here.

In the central section of the Great Barrier Reef, the continental shelf widens, making a broad shallow area for patch reefs and coral-fringed islands. This is the most accessible part of the reef, and tourism and development have had a considerable effect.

Farther south, the continental shelf continues to widen. Submerged reefs run along its edge, and behind them, patch reefs cover vast areas, with "complex lagoons and deep, tortuous channels." One, the Pompey Complex, is 120 miles long and 9 miles wide. Despite its southern position and cooler waters, this part of the Great Barrier Reef has a high diversity of corals and fish species.

At the system's southern end, the reefs grow gradually smaller and less connected as the continental shelf narrows and water temperatures slip closer toward the minimum required by corals.

Thus the Great Barrier Reef is neither a continuous barrier nor a single, interconnected structure or organism. Nonetheless, its name seems right. Like a barrier reef, its outer portions divert ocean currents and intercept ocean swells. And its greatness is beyond question; it is one of the planet's great geological and biological features, with a history stretching back twenty million years in its oldest northern portions. For the underwater explorer, who thrills in its diverse coral reef ecosystem second only in diversity to that of the tropical rain forest, does it really matter what we call this Underwater Wonder of the World?

Marbled scorpionfish are quite common on the Great Barrier Reef. Like other scorpionfish, these solitary predators have adopted pigments and scale textures that match their surroundings.

. . . .

ON THE THIRD DAY OF DIVING, WHEN OUR JET LAG FINALLY BEGINS to fade, naturalist Rob Gomersall, our dive guide, takes me to explore Osprey Reef in the Coral Sea. Diving here is quite different from my earlier Great Barrier Reef experiences. The reef is badly damaged, not by careless divers, but by a cyclone that devastated the area in 1988. On the sea floor, huge coral boulders lay overturned, torn from the reef by the force of the waves. Colorless and comparatively lifeless, this reef is a vivid reminder of just how fickle nature can be.

But I wonder if the cyclone should be blamed entirely on natural causes. Some scientists suggest that global warming is the reason behind a recent increase in the number of cyclones or hurricanes. Perhaps the damage to this reef is one more proof that each of us needs to take better care of this earth of ours.

These damselfish nest among the sturdy branches of staghorn formations. The male cleans off a piece of dead coral, then dances to attract a mate. The female deposits her eggs on the coral, after which the male drives her away. He then guards the eggs for four days until they hatch. Damselfish take about a year to mature and may live for three to five years.

Facing The decorator crab has developed a unique way to deceive predators. Clipping off tiny pieces of sponge and algae, the crustacean attaches them to its shell, becoming almost invisible on the reef—until it moves across a different background.

Another reason for the reef's demise may be the population explosion among crown-of-thorns starfish, called the devilfish by the Japanese. These predators, one of the largest starfish in the ocean, feed frantically on corals, sometimes in packs of thousands, leaving a path of destruction as they crawl over the reef. In their wake, algae swiftly dominate the dead reef surfaces, delaying repopulation by new corals.

AIMS scientists have noted that between the two major outbreaks of crown-of-thorn starfish, in the early and mid-1960s and early 1980s, the coral communities took almost ten years to regenerate—only to be hit in many cases by the second devastating infestation.

Just as some people believe that humans may be responsible for the recent increase in cyclones, some students of the reef have asked whether these two plagues of the crown-of-thorns starfish may have occurred because of human activity. Have we removed a natural predator, or upset the natural balance of the reef in a way that leaves the starfish population free to expand wildly?

The question is difficult to answer. The immense area of the Great Barrier Reef, and the lack of detailed research before 1960, leaves scientists without much information about the history of crown-of-thorns starfish infestations. Nobody can say for sure whether what is happening now has happened before.

Scientists do know that this starfish has the ability to reproduce in huge numbers if conditions are right. An average female crown-of-thorns can produce thirty million eggs a year. If a few thousand are fertilized and a few hundred reach maturity, and the same results are achieved by thousands of other crown-of-thorns, a population explosion is under way. What triggers these episodes is now being researched.

The answers will be important for coral reef biologists. In the past, reefs were regarded as extremely stable environments, and it was assumed that their diversity depended on this stability. But if researchers discover that the incredible diversity of the Great Barrier Reef is sustained despite frequent onslaughts of the crown-of-thorns starfish over large areas, then the concept of reef stability may need to be reexamined.

The red firefish hunts by using its stinging fins to corral smaller fish into holes on the reef where they can't escape. Once its prey is cornered, the firefish sucks it into its mouth. Divers who see this near the surface will hear a sound like a loud clap of hands.

Today the coral rubble at Osprey Reef isn't completely barren. As I explore, I happen upon a most interesting creature lying on the sea floor, a foot-long leopard sea cucumber, which, as you may imagine, looks like a cucumber with spots. When touched, this creature, the sea's underwater vacuum cleaner, spews out much of its internal organs, in sticky strands coated with a toxic substance that looks like white spaghetti. This defensive technique, designed to confuse and entangle attackers, certainly works in my case. I keep my distance—as one should with all reef creatures. I've read, however, that in much of Asia, the sea cucumber, also known as bêche-de-mer, is a delicacy.

.   .   .   .

TODAY THERE IS NO WIND, NO CLOUD COVER, NO SURFACE CURRENT. In calm seas, I can look forward to what divers dream of: great underwater visibility. Here, at a bommie called Steve's Spot, I'm preparing for "macro heaven," a name photographers give to dive sites thriving with tiny creatures.

At forty feet, I find a fairy-tale environment filled with literally hundreds of small animals I want to photograph. But since I'm limited to thirty-six frames, I must choose my subjects carefully.

Feather starfish are everywhere. In fact, there is a higher concentration of feather stars on the Great Barrier Reef than anywhere else in the world. If you took two hundred or so delicate, three-inch-long feathers and joined them together at a base of a dozen or so legs, you'd have a creation that looks somewhat like these fragile plankton feeders, whose arms break off at the slightest touch.

Throughout my dive I take close-up pictures of bottom-dwelling and sedentary fishes, including different species of lionfish and stonefish. These fishes are not equipped with a swim bladder (used to maintain buoyancy), as are butterflyfish, angelfish, and other free-swimming fishes. Instead of darting away at lightning speed into the safety of the reef, lionfish and stonefish employ another defensive measure: venomous dorsal spines that inflict excruciating pain on predators—fish and human alike.

Beautiful as it may seem, the reef is not a fantasy land where nothing bad can happen. The sting of the stonefish is sometimes fatal to humans, and the

The sea cucumber has evolved two remarkable ways to distract predators. Sometimes the bottom-dwelling creature releases the sticky white stringlike substance shown here. As the mass entangles the attacker, the sea cucumber moves to a safer location. In other cases, the sea cucumber eviscerates its digestive and respiratory tracts and gonads. The predator gets an easy meal, and the sea cucumber crawls away and regenerates these organs.

Nudibranchs are marine relatives of the far less colorful terrestrial slug. Dozens of species inhabit the Great Barrier Reef. Researchers describe them as elusive, if not rare, and continue to discover new species every year.

Like other anemonefish, the false-skunk-striped anemonefish lays its clutch of eggs near anemones for protection. It tends the eggs by fanning oxygen-rich water over them—an essential step if they are to hatch.

Great Barrier Reef hosts dozens of other highly poisonous creatures—mollusks such as the geographer cone and the tulip cone, fishes such as catfish, stingrays, and rabbitfish, and stationary creatures from fire corals to fire anemones.

On a more pleasant note, I find several different species of nudibranchs, delicate, sluglike animals that glide over the corals in search of detritus and bacteria. Although these animals are related to snails, one couldn't tell from their looks. They have no shell and their gills are fully exposed. Their name, in fact, means "naked gills" in Greek. Perhaps the most beautiful animals on the reef, nudibranchs range in size from less than an inch to more than a foot long. In my travels, I've seen nudibranchs in many colors. Here, I find a new species to add to my life list of marine creatures: a three-inch-long, bright yellow *notodoris negastima*, quite uncommon in this area.

My air supply is running low, so I begin my ascent along the wall of the bommie, hoping for a really great picture to conclude this dive. Soon I notice a pair of clownfish darting in and around a huge sea anemone. I've seen this sight before, but these fish seem unusually excited. Then I see why: they are protecting and nurturing a tiny patch of eggs attached to the rocky substrate just underneath the fleshy sea anemone.

One fish darts out to ward me off—an intruder hundreds of times its size! The other blows a fresh supply of sea water over the eggs, providing a steady stream of oxygen-rich water for the developing clownfish. I click off my last frames, feeling frustrated that I can't spend more time documenting the scene and hoping I get a good exposure of the fast-paced action.

Frustration is a feeling that has been with me for the past few days. I've taken hundreds of pictures of the flora and fauna of the Great Barrier Reef; only a few will be selected for the book. Can I fully tell the story of this enchanting region? The answer, unfortunately, is no. In my week here, I've only visited a few reefs in the northern section. A thousand miles of reef complexes stretch away to the south, and on this visit, I won't see any of them.

Talking with people who work full-time on the reef, the scientists, park managers, and dive guides who know this coral world best, I learn that my frustration is hardly unique. For scientists, the Great Barrier Reef raises thousands

of biological questions—big theoretical issues such as the dispersal of species over time and space, and smaller but equally important inquiries into the behavior and life cycle of individual species and the exact nature of their relation to the large ecological web. Scientists are also concerned about the future of the reef. The inventions of helicopters, outboard motors, and scuba gear have opened the most remote parts of the reef to human exploration and exploitation. Before the

Bottom-dwelling sea lilies or feather stars use their hundreds of arms to trap plankton. A smaller number of limbs gives this relative of sea stars the mobility to walk from daytime hiding places to nocturnal perches amid plankton-rich currents. Fossilized remains reveal that ancestors of the feather star lived more than five hundred million years ago.

1950s, most of the reef was inaccessible; now it is all accessible, and scientists must include measurements of human impact in their long list of studies to pursue. There's far from enough time or money for all the things researchers would like to know about the Great Barrier Reef.

The people of Australia have long recognized what an amazing treasure lies along their northeast coast. Some say that the Australian environmental movement began in 1967, when a mining company sought permission to extract limestone for agriculture from the reef. Opposition quickly arose, and the mining operations never began. In 1969, the government gave oil ventures permission to prospect for oil on the Great Barrier Reef. Conservationists fought this too, and a 1975 law banned that activity.

In 1975, the Australian Parliament decided on a more thoroughgoing plan to guarantee the future of the reef, naming the whole complex a national park and setting up the Great Barrier Reef Park Management Authority. Slowly, the reef is being mapped, analyzed, and classified in zones of usage. Most areas are assigned to controlled general use, where tourism, commercial fishing, and sport fishing are allowed. Smaller portions are designated as marine national parks, where all activities are far more strictly limited. And about 1 percent of the park is refuge, where no human activities other than scientific research are permitted.

For everyone who thinks about the future of the reef, tourism is the main topic of discussion. Every year, more tourists come, more hotels are built, more boats filled with eager divers venture out to remote spots. Perhaps reefs can adapt to cyclones or the crown-of-thorns starfish, enemies they've faced over the eons. But no one knows how well they can adapt to human beings. The hope is that as we learn more about the Great Barrier Reef, we will also learn how to enjoy it without doing harm.

Countless numbers of shrimps and other invertebrates can be found on the Great Barrier Reef. Red reef shrimp and other species are more easily seen and photographed during night dives, when the animals leave the safety of hiding places in the reef in search for food.

Facing *Clownfish, which grow to a length of two or three inches, are some of the most aggressive fish in the sea. These tiny creatures "attack" divers in hopes of driving them away from their home—the stinging tentacles of the sea anemone, which trap other fishes but have little or no effect on the clownfish.*

Deep Ocean Vents

I FIRST LEARNED ABOUT DEEP OCEAN VENTS IN A 1979 ISSUE OF *National Geographic*. Emory Kristof's photographs amazed me: six-foot-long tube worms, foot-long albino clams, sightless crabs, and dandelion-shaped siphonophores living thousands of feet below the ocean's surface in a sunless seascape of hydrothermal springs, lava lakes, and mineral-rich pillar formations.

Since then, I've often dreamed of exploring and photographing deep ocean vents. They clearly belong among the Seven Underwater Wonders of the World. But the cost of time in a manned submersible (about twenty thousand dollars a day in 1991) put them far beyond my reach.

However, thanks to Emory and three other friends—Dr. Dan Fornari of the Lamont-Doherty Geological Observatory at Columbia University, Dr. Rachel Haymon of the University of California at Santa Barbara, and Al Giddings of Ocean Images, Inc.—I'm able to share with you unique photographs from the world's least accessible underwater wonder. In addition, Emory describes why the vents are an important discovery, while Dan and Rachel tell the story of a visit in *Alvin* to an active vent and survey some remarkable aspects of vent biology and geology.

The Deep Ocean Vent Environment

by Emory Kristof

THE SIGHTS I SHARED WITH GEOLOGIST KATHY CRANE AND PILOT Dudley Foster from the portholes of the deep sea submersible *Alvin* in 1979 were not of the earth I thought I knew. Eighty-five hundred feet below the surface in the Galápagos Rift, we were surrounded by large, white-stalked, red-headed tube worms swaying gently in hot volcanic vent water.

Facing *Since 1964, the submersible* Alvin *has enabled American researchers to visit the depths of the ocean. Equipped with televisions, still and stereo cameras, equipment trays, a water sampler, an acoustic velocity meter, and a mud grabber,* Alvin *can descend to twelve thousand eight hundred feet below the surface. She is named for Allyn Vine, a scientist at Woods Hole Oceanographic Institute who advocated the usefulness of deep-sea submersibles and helped design* Alvin. *Emory Kristof, a staff photographer for* National Geographic, *has been lucky enough to accompany research dives in* Alvin *in several of the world's oceans. (© National Geographic Society)*

Red-tipped giant tube worms, some twelve feet in length, exist without sunlight around deep ocean vents thousands of feet below the surface. A pigment in the tube worm's blood contributes the bright red color, seen with the aid of a powerful underwater flash. Not seen in this photograph are the animal's three hundred thousand tiny tentacles, which absorb ogygen as well as bacteria from water near the vents. (© Al Giddings, Ocean Images, Inc.)

Before the late 1970s, no one suspected that large numbers of animals lived anywhere in the deep ocean, least of all around geothermal vents. Photographs from remote cameras confirmed that various species of urchins, brittle stars, and sea slugs exist on the seafloor between four thousand and eighteen thousand feet, and that they do a rather efficient job of mixing up and homogenizing the sediments on the seafloor. Besides these inhabitants, however, the abyss was thought to be a fairly lifeless, monotonous region of the planet.

All that changed with the discovery of tube worms on the Galápagos Rift in 1977 and of other animal colonies around superheated "black smoker" vents on the crest of the East Pacific Rise at 21° N, less than two hundred nautical miles off the western coast of Mexico, in 1979.

The geologists and geophysicists who found these populated vents weren't looking for evidence of life in the abyss. They were participating in an international, interdisciplinary study of the largest feature on the face of our planet, the Mid-Ocean Ridge, a forty-thousand-mile-long volcanic mountain range and rift system that snakes across the bottom of the world's oceans like the stitching on a baseball. Like stitching, the ridge system marks where pieces of the earth's outer layer, its lithospheric plates, meet.

Where these plates meet, along the crest of the Mid-Ocean Ridge, volcanic and magmatic forces continually form new crust as the plates spread apart. In some places, such as the Mid-Atlantic Ridge, the spreading is slow, about an inch per year. By contrast, on the southern end of the East Pacific Rise, spreading occurs at a rate of eight inches per year. Probing the ridge crests with manned submersibles and remote sensors, researchers sought out places on the ridge crest where geological processes are most intense, hoping to gather new insights into plate tectonics.

They weren't disappointed. The discovery of deep ocean hydrothermal vents, and their importance for understanding many geologic and oceanographic processes, have completely redefined marine scientists' understanding of earth-forming processes. The chemistry of deep ocean vent waters has helped clear up a great mystery about the composition of ocean water. Moreover, for the first time, biologists can study a food chain that functions without sunlight. Both phenomena depend entirely upon the complex workings of deep ocean vents.

These vents form as a result of the volcanic and tectonic processes that accompany seafloor spreading. As the plates spread apart, cracks and faults are created through which magma erupts. Usually the lava flows slowly and cold seawater solidifies it in bulbous pillow shapes, but sometimes it rushes up with such ferocity that it spreads out in lakes streaked with swirls and coils.

Over time, more cracks appear in the surface of the volcanic flows, and seawater, under a pressure of thousands of pounds per square inch, flows down toward the magma, which heats it to temperatures as great as 660° F. As the water heats, it releases magnesium and other minerals and picks up heavier ones such as compounds of iron, zinc, copper, manganese, silver, and lead. The hot mineral-laden water then rises through other openings in the crust, reentering the ocean through the vents.

It used to be thought that all minerals were derived from the erosion of the continents and were carried into the ocean by rivers. But the relative quantities of minerals in seawater did not correspond to their quantities in rivers. Deep ocean vent circulation supplies the missing pieces of the equation. Some scientists now think that all the world's seawater circulates through the crust once every ten to twenty million years.

As the hot water circulates through the basaltic rocks, hydrogen sulfide is formed. This compound plays a key role in the unique food chain of the vent communities. Certain marine bacteria metabolize this mineral in a process called chemosynthesis. The microbes multiply rapidly in the sulfide-rich environment of the vents, creating a food supply for other marine creatures far from the sun's reach.

Before the vent communities were discovered, biologists believed that only sunlight, through photosynthesis, could support so populous an ecosystem. Now some speculate that life on earth could have originated in deep ocean vents.

The biology of the vents has produced many surprises and raised many questions. These communities are tiny oases of life in the middle of vast marine deserts. Once the animals are there, they feed on the enormous supply of bacteria. But how do these creatures find the vents?

Marine biologist Frederick Grassle is accustomed to the tight working quarters of Alvin. From the safety of the cabin, he can study the unique creatures of the deep, which are supported by a food chain like no other on earth. Until the vent communities were discovered, biologists believed all life on earth ultimately depended on sunlight. But life at the vents begins with a specialized bacteria that metabolizes hydrogen sulfide. The bacteria, in turn, becomes food for the other creatures of the vent community. (© Al Giddings, Ocean Images, Inc.)

Mussels and crabs are among the creatures which thrive on bacteria at deep sea vents. Evolving in an utterly lightless habitat, the blind white crabs have no need of sight or of bodily pigmentation. (© Woods Hole Oceanographic Institution)

Although a variety of creatures surround each vent, a particular animal usually predominates, leading biologists to speculate that the first species to arrive tends to outnumber later arrivals. At the Galápagos Rift, hundreds of fast-growing twelve-inch clams mark one site, known as Clambake, while the so-called Rose Garden features the largest concentration of tube worms. At the Trans-Atlantic Geothermal Vent south of the Azores, thousands of members of a newly discovered species of shrimp swarm through black clouds of metal-rich water. At other sites around the world, mussels, snails, octocorals, and polychaete and serpulid worms are the prime established settlers.

Many of the vent creatures may have long larval states—a topic which deep ocean vent biologists are studying intensively. Perhaps, like plant seeds riding the wind, they drift great distances in ocean currents before they encounter a vent. If this is how animals populate the vents, it is a strategy fraught with obstacles. Perhaps a better explanation will be found. The hope of answering these and other questions is the siren song which lures scientists to continue their exploration of the deep ocean vents.

"Black smokers," plumes of metal-rich, superheated water, have been documented at numerous vents on the Mid-Ocean Ridge crest. They get their hellish look from the iron, zinc, copper, manganese, silver, lead, and other minerals picked up as the 660° F water circulates through fissures in the oceanic crust. (© Woods Hole Oceanographic Institution)

LIFE AMONG THE VENTS

Dr. Daniel J. Fornari and Dr. Rachel M. Haymon

THE DARKNESS SEEMS NEARLY COMPLETE AS WE DESCEND THROUGH the blue-black water. Visible light reaches down only a thousand feet or so in the ocean; below about thirteen hundred feet, all vestiges of measurable light disappear. From the portholes of the submersible *Alvin*, we see a rich mass of phosphorescent particles swirling around the descending sub. Tiny animals with too many legs, antennae, and tentacles brighten the abyss with patterns like miniature psychedelic marquees. Only the hum of the sub's equipment disturbs the scene, one that has not changed much since life began in the oceans.

Suddenly, the pilot turns on the exterior lights in preparation for the bottom approach, and what was a serene if somewhat bizarre circle of deep black ocean becomes a blinding opaque haze. Inside *Alvin*, activity heightens in the pressure sphere, the titanium-steel housing where three of us crouch throughout the ten-hour dive. We prime our cameras and other recording devices to begin collecting data. We scan the depths below the sub with our bottom-looking

The abundance of life at vent sites draws some familiar creatures. Comfortable under the pressure of 250 atmospheres, an octopus scavenges for a meal of crabs and mussels. (© Jack R. Dymond, Oregon State University)

sonar and the area around the sub with the forward-looking sonar to make sure that we do not crash into a rock wall or volcanic outcrop, and we call the ship on the underwater sonar telephone to let them know that we are making a bottom approach.

Now the seafloor lies sixty feet below the sub. Vague forms appear at the outer edges of the cone of light cast by our external lamps, shapes that recall the volcanic landscapes of Hawaii or Iceland. The terrain is a jumble of intertwined volcanic lava flows in shapes that range from sofa-sized bolsters to spherical pillows nearly three feet in diameter, surrounded by toothpastelike tubes that seem to lead in all directions. The lava surfaces are jet black, and their glassy crusts glitter under the sub's lights. Off to the side, we also see a wide beige expanse as flat as a pool table. Then our propellers stir up a dusting of calcium carbonate sediment, revealing the black surface of a lava lake, which congealed in the near-freezing bottom water shortly after the eruption that created it.

The appearance of a few small, fierce-looking Galatheid crabs heralds a change in the character of the bottom, which is mirrored by the heightened anticipation of the pilot and crew of *Alvin*. The sediment is now a light ochre, and we see many more crabs around the volcanic outcrops. "We just got to the suburbs," someone says to break the tension. Now there are crabs everywhere, as well as large white clams and brownish mussels in every crevice. Some lava outcrops appear to have grown short white beards—the tube stalks of serpulid worms.

Rachel notes that the water looks milky in the distance, and as she finishes the sentence, we see the large red-tipped tube worm colonies. Just behind them, a chimney belches a stream of black fluid into the water. We have arrived at an example of the world's most recently discovered natural wonder—a deep ocean hydrothermal vent.

At a "black smoker" vent such as this, super-heated water spews out at temperatures over 660 °F. (Vent fluid temperatures can far exceed water's boiling point because of the high restraining pressure that exists thousands of feet below the surface of the sea.) Only on the crest of the Mid-Ocean Ridge do these high-temperature, deep-ocean hydrothermal vents appear. Three to four thousand feet below the ridge crest, a magma chamber stores the molten rock between

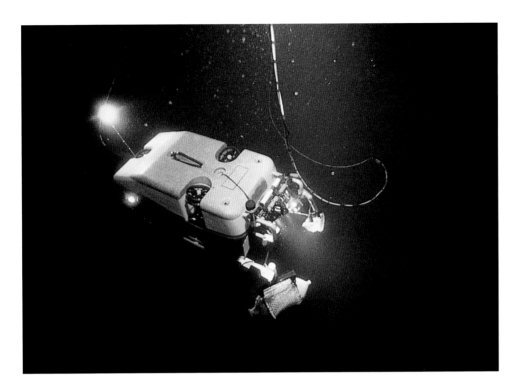

Jason, *a remotely operated underwater research vehicle, documented vents in the Mediterranean Sea in 1989. A team of scientists and photographers guides* Jason's *seafloor explorations from the cabin of a research vessel on the surface. While manned submersibles such as* Alvin, *France's* Cyana, *and Russia's* Pisces *allow scientists to visit deep-sea sites in person,* Jason *and other remotely operated vehicles are better suited to large-scale surveys of the ocean floor.* (© *Quest Group, Ltd., Woods Hole Oceanographic Institution)*

eruptions. Cold water flows into cracks and fissures in the volcanically heated crustal rocks, and the deeper it penetrates into rocks heated by the flow of magma, the hotter it becomes. The high temperatures enable the water to leach various compounds from the minerals in the volcanic rocks, including metals and the bacteria-supporting sulfides, and deposit other compounds in small veins and cracks in the rocks.

After finding other paths back up to the seafloor, the hydrothermal fluid that erupts from the vents may be warm (37°–68° F), intermediate (175°–400° F), or very hot (more than 660° F), depending on the magmatic activity and crustal permeability of a particular segment of the Mid-Ocean Ridge crest. Vents in different parts of a three- to six-mile crest section may have different temperatures and slightly different hydrothermal fluid compositions depending on the plumbing system of cracks and fissures that transports the fluids through the volcanic crust, the chemical composition of the lavas, and the local heat transport processes.

Deep ocean hydrothermal vents often form tall, narrow chimneys on the lava flow surfaces of ridge crests. Individual chimneys can be as much as eighty feet high and a dozen feet in diameter. At well-established vent complexes such as those on the Mid-Atlantic Ridge, the height from the base of a hydrothermal mineral mound to the top of a cluster of chimneys may exceed 160 feet. Both the spirelike chimneys and the broad, low mounds that sometimes underlie them are composed of iron-, zinc-, and copper-rich sulfides, all of which have precipitated from the vents' mineral-rich hydrothermal solutions.

Mineral precipitation occurs quickly around the chimney opening because of the extreme temperature difference between the superheated vent fluid and the near-freezing water of the abyss. Layers of crystals, including pyrite, chalcopyrite, and sphalerite, form concentric rings around the opening of the vent, each layer thickening the walls and increasing the height of the chimney for as long as the flow of hot fluid continues. Sometimes the precipitation of minerals is so rapid that it chokes off the outlet of the vent. The percentage of metallic minerals is often very high in hydrothermal deposits; some of the earliest copper mines of the Greeks on Cyprus were formed a hundred million years ago by hydrothermal processes on an ancient Mid-Ocean Ridge crest.

Hidden a mile or two below the surface of the sea, most of the world's deep ocean vent areas have yet to be discovered. Since the first vents were found fifteen years ago, only 1 percent of the Mid-Ocean Ridge crest has been surveyed. An even smaller area has been surveyed in sufficient detail to find hydrothermal vents. Nonetheless, researchers have located vents on the Mid-Atlantic Ridge, the East Pacific Rise, the Galápagos Rift, the Juan de Fuca Ridge (just off the coast of Oregon and Washington), and on isolated segments of the Mid-Ocean Ridge in the western Pacific.

The biology of deep ocean vents has revolutionized the biological sciences, as researchers analyze the significance of a food chain which evolved without the benefit of photosynthesis. The sulfide-eating bacteria upon which the vents' clams, mussels, crabs, and worms feed may be a clue to how life formed billions of years ago.

The last decade of research has shown that hydrothermal vents play a significant role in the composition of our hydrosphere and biosphere, but most of our discoveries in the geology, chemistry, and biology of the vents remain to be made.

Facing Getting Alvin *to a vent site requires teamwork. On this launch near the Galápagos Islands, a dozen people help to get the manned submersible into the water. Funding for research in* Alvin *comes from the National Science Foundation, the National Oceanic and Atmospheric Administration, the Office of Naval Research, and other sources. (© Emory Kristof, National Geographic Society)*

PALAU

"THE CRADLE OF DIVERSITY"—THAT'S WHAT MANY MARINE BIOLOGISTS call the waters of the tropical western Pacific within the triangle formed by Guam, the Philippines, and New Guinea. Why the distinctive designation? Simply put, these waters are home to the richest collection of marine plants and animals in the world. Near the center of this triangle—7° 30' north latitude and 134° 30' east longitude—lies the archipelago of Palau, a string of islands around which, it is said, more different marine species can be found than anywhere else on earth.

Robert E. Johannes, an Australian marine biologist who nominated Palau at our Seven Wonders meeting in August 1989, calls the archipelago his marine Mecca, the destination of choice for his scientific pilgrimages. Writing to me about Palau, he gave some telling examples of what diversity means there. "A few years ago," he writes,

> two colleagues of mine collected one genus and thirteen species of fish new to science while diving on the reefs near Palau's district center. The magnitude of this achievement can be judged by the fact that each year, only seventy-five to a hundred new species of fish are discovered worldwide by researchers in marine and fresh water.
>
> Another colleague recorded 163 species of corals in less than twenty feet of water on a 400-yard-long survey in Palau's main harbor—twice as many species as have been found in the entire Caribbean to a depth of 300 feet.

Using seagrass as an example, Johannes describes how the diversity of tropical marine life diminishes as one moves away from Palau:

> Palau has nine species of seagrass. Yap, 350 miles away, has only five. Still further to the west is Truk, with four. Ponape is next, with two. And by the time we get to the Marshall Islands, roughly fifteen hundred miles away in the central Pacific, we find only one.

The sea fans of Palau grow to enormous sizes, some reaching a height of ten feet. Sea fans—and all other soft corals—are extremely delicate and fragile, and should never be handled.

Facing *True to their name, hawkfish scout for prey from vantage points on the reef, striking quickly when they spot a likely victim. Over time, they have evolved strong pectoral fins that enable them to wait on their perches without making telltale movements.*

Above and facing *The lush islands of Palau began to form when geological forces raised submerged coral-covered volcanic mountains above the water. Exposed to air, the corals died, leaving only their porous limestone skeletons. As birds, currents, and wind brought seeds from other lands, a wide variety of tropical vegetation took root. New coral reefs grew in the shallows surrounding the islands. Today, these reefs abound with seven hundred species of coral and more than fifteen hundred species of fish.*

Johannes observes that Palau's high marine diversity is matched by a wide range of habitats, "a greater variety than can be found in any similar-sized area in the world." Within the approximately ninety-mile-long archipelago, there are fringing reefs, lagoons, patch reefs, barrier reefs, mangrove swamps, and seagrass beds—not to mention the islands' three dozen marine lakes. Each of these secluded salt-water pools, surrounded by dense jungle, harbors a unique assortment of marine species, "forms," writes Johannes, "which could only have evolved in such sheltered conditions." Even during typhoons, the waters of these lakes are completely protected.

The marine lakes are complemented by Palau's numerous marine coves, which evolved geologically as erosion rejoined some lakes to the sea—often very tenuously. "Some have a single entrance so narrow that visitors can only get in at high tide in a small boat," writes Johannes. In one, marine biologists were astonished to find all seven of the world's species of giant clams.

Today, Palau is virtually unspoiled. It's not easy to get to, so the impact of tourism is still minimal. The Micronesian culture of the Palauans is relatively intact. Palauans respect and care for their surroundings, and their traditional customs are intended to prevent the depletion of the fishes and other marine life they depend on.

Yet Johannes tells me that this remote tropical archipelago, this Mecca of marine biology, is threatened. The United States, which oversees and protects the islands, wants to dredge some of Palau's lagoons and reefs for a major new submarine base.

The impact of this massive military project would not be limited to the areas destroyed by dredging. Palau's small population, fourteen thousand in all, would feel the cultural impact of the numerous American servicemen who would be stationed there to tend the submarines. American dollars would alter the local economy, eroding centuries-old values and customs. As we have seen in remote places around the world, a decline in concern for the environment often follows the influx of foreign ideas and money and the loss of traditions.

In Palau, where people have lived in harmony for centuries with one of the richest marine environments in the world, the degradation of that environ-

ment would be a tragic and incalculable loss. "Palau's reefs and lagoons may be resilient," writes Johannes, "but the damage caused by a major military installation could take centuries for nature to repair." He asks, "Will western legalisms and economic blandishments prevail? Or will Palau's ancient laws—which predate the earliest western efforts at conservation by many centuries—help to halt the U.S. military in its drive to turn Palau into a Pacific stronghold?"

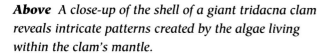

Above *A close-up of the shell of a giant tridacna clam reveals intricate patterns created by the algae living within the clam's mantle.*

Right *Giant clams have existed on earth for more than two hundred million years. In Palau, these mollusks can grow to be five feet long and weigh several hundred pounds. Islanders have used giant clam shells for hundreds of years as a tool for crushing taro and tapioca, and the meat is considered a delicacy, especially when eaten raw. These tridacna clams were photographed in the shallow-water Palau Clam Farm, where a successful breeding program is under way.*

Facing *Virtually every inch of the reefs in Palau teems with life. In a one-square-foot section, a diver may find a soft coral colony, spotted sea squirts, encrusting sponges, and several different species of colorful algae.*

DIVING AT NGEMELIS ISLAND, I'M IMMEDIATELY CONVINCED THAT I am indeed at the center of marine diversity.

On this mile-long island, terrestrial wonders include steep slopes clothed in dense jungle and dramatic, wave-carved limestone formations that punctuate the beaches. But for me, the true wonders lie beneath the water.

The reef starts almost immediately under the surface. Even from our boat, I can see the dazzling colors, variety, and diversity of the thriving reef below.

Once underwater, at a depth of only ten feet, I'm overwhelmed. In all my dives on our Seven Wonders expeditions, I have never seen this much marine life in one place at one time. Green, yellow, red, and orange branching soft corals dot the shallow reef plateau. Branching hard corals are here too, in shades of green, brown, yellow, and blue.

Nestled among the corals are giant clams, gracefully exposing their almost luminescent mantles in the sunlight; sea lilies swaying in the light current, trapping floating plankton; and sea anemones, surrounded by clownfish and dominofish, which live in these animals' tentacles. Scattered across the seascape, a variety of brittle starfish and cushion starfish add their shapes to the bewildering mosaic.

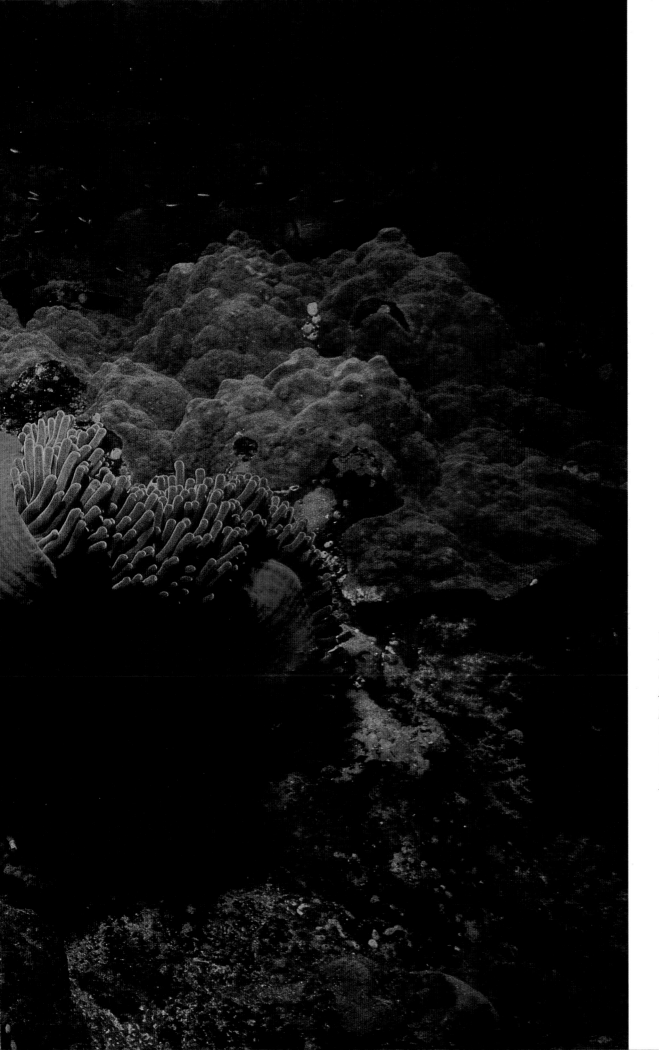

With a diameter of three feet or more,
Palau's sea anemones are among the
largest in the Indo-Pacific. False skunk-
striped clownfish and other anemone
fishes, often seen in pairs, hover about the
tentacles of these giant cnidarians. Like
their relatives the corals, sea anemones use
their tentacles to draw prey into their
stomach cavities.

For centuries, Palauans have depended on reef creatures for food, and their traditions include methods for preventing overharvesting or other damage to the reef. Westerners lucky enough to visit this remote archipelago will find that thriving reef communities begin just below the surface—a snorkeler's dream.

Facing A close look at a soft coral branch reveals a tiny spider crab. This camouflaged crustacean may spend its life on one coral colony.

*Surrounded by algae, **facing**, the crocodilefish virtually disappears; even the eyes of this masterful ambush artist are camouflaged. On a coral colony, **right**, the bottom-dwelling predator becomes more visible. The crocodilefish's eyes, set practically on top of its head, are well suited to watching for prey in the waters above.*

Under ledges, crocodilefish and scorpionfish, as well as less menacing shrimps, crabs, and lobsters, lie in wait for their next meal. Hovering above the reef, I observe the underwater ballet of angelfish, parrotfish, big eyes, pufferfish, and yellowtails. All this in the first five minutes of my dive!

As I continue my exploration, seeing more and more species every inch of the way, I stop photographing and simply gaze in awe. An interesting realization comes to me: of all my dives in the past eleven years, this is the best.

Ascending toward reality, or perhaps just a different reality, I'm comforted in knowing that the Palau government favors conservation and is implementing programs to protect the area for future generations. It would be disastrous if this pristine environment suffered the fate of so many underwater locales, its beauty lost to humankind forever. Thinking of what Johannes has told me about U.S. plans for Palau, I can only hope that getting the word out through books, newspapers, and television programs will help preserve the archipelago.

. . . .

WE ARE CLIMBING THROUGH DENSE JUNGLE ON EIL MALK, A TWO-mile-long island just southeast of Koror, the main island. I'm carrying two heavy underwater cameras, each equipped with two strobes. Susan, three months preg-

nant with our first child, follows me with two large net bags containing our snorkel gear.

On the jagged, winding path, lined with sharp coral skeletons, we must step carefully, as our hands aren't free to steady ourselves. Perhaps it's best this way: we have been told there are poisonous trees on the island that cause severe allergic reactions when touched. Cautiously, we move on.

The temperature is only in the eighties, but the humidity is so high that my glasses are fogging up. At the top of a ridge, encircled by hanging vines, we take a breath, enjoying the sounds of the forest—the hum of insects, the call of the long-tailed frigate birds, and the screech of brilliantly plumed parrots.

After catching our breath, we begin our descent. Slowly, the marine lake comes into view through the forest. I wonder if we'll see the salt-water crocodile we've heard about?

Five minutes later, we reach the lake shore, lined with thick mangroves and tall trees. After a brief search, we find a narrow channel and put on our snorkel gear.

Susan, twenty yards in front of me, spots a lone jellyfish. She calls me through her snorkel, and I swim over to see this wonderfully simple melon-sized creature. After taking several photographs, I again hear Susan calling me. "There are hundreds over here," she says. When I join her, I see that we are completely surrounded by jellyfish. They are swimming in all directions, bumping into us as well as each other. We are "getting slimed," as those who know the lake say.

Getting slimed actually feels kind of nice in the warm salt water—like being massaged with warm Jell-o. In any event, it's better than getting stung. But there's no need to worry. Sealed off from open-water predators, these jellyfish lost their stinging capability long ago.

Suddenly, it begins to rain—hard, fast, almost horizontal. Susan says, "Let's go," but from the center of the almost perfectly oval lake, the mangroves form a solid wall of green. We don't know which direction to swim. We are lost in Jellyfish Lake. Enjoying the moment, we laugh at how this will sound to our friends back home.

The rain, like most tropical squalls, is over within minutes. The hot sun

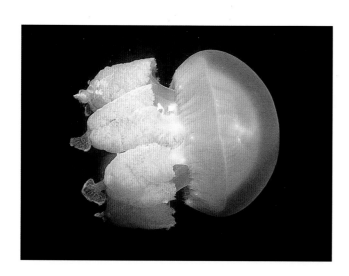

When Palau's Jellyfish Lake was sealed off from the sea eons ago, jellyfish were trapped in the warm, shallow water. Numbering in the thousands, these cnidarians pose no threat to swimmers. Living in a habitat without predators, they long ago lost their stinging capability.

All of the wonders of Palau are not under water. Topside, the remains of extinct coral reefs create striking formations throughout the archipelago's twenty-six islands.

Above and facing *During the day, schools of chromis feed on plankton above the reef; at night, they doze among the branches of soft corals or in other safe places. Like many reef creatures, these tiny fish are very territorial and can probably be found in the same spot each night.*

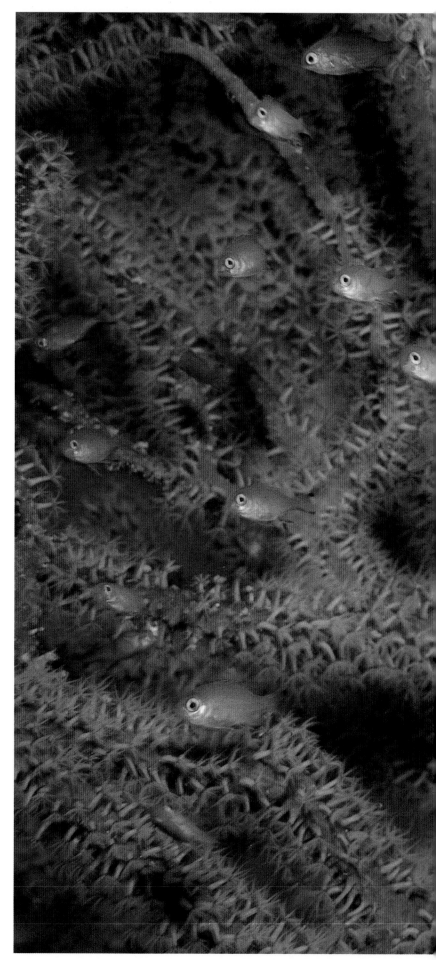

is out again and a rainbow sweeps across the sky. Now we can see our mangrove channel, our path back to civilization—in this case, a tiny country with a thousand times more water than land.

.   .   .   .

HARD RAIN AND STRONG WIND STRIKE AGAIN JUST AS WE ARE ABOUT to explore a site called Blue Corner. Gearing up for the dive, trying to maintain my balance on the rocking dive boat, I'm reminded of the old commando movies where the paratroopers jump out of the plane in rapid succession. The boat captain has just told us, six fairly skilled divers, that we must enter the water one right after the other. If we delay, he says, the strong current will separate us too much, making the diver pickup after the dive lengthy and difficult.

After we dive in and begin our descent along the wall, the two-knot current whisks us away. I've done drift dives before, but none in so strong a current. I want to stop and photograph the sea fans, sea whips, soft corals, fishes, and sea anemones on the wall, but it's no use.

Fifteen minutes into the dive, we reach Blue Corner. The current is not as strong here, so I stabilize myself by holding onto a piece of dead coral. As always, I'm extremely careful not to touch the fragile living coral.

Above me a school of perhaps several hundred barracudas circles like vultures over a road kill. Looking away from the reef, I see a dozen or so reef sharks on patrol, eyeing the strange intruders who have entered their hunting territory. To my left, there are huge schools of triggerfish and butterflyfish swimming in tight formation. Perhaps they believe there is safety in numbers. And to my right, a school of jacks almost completely blocks the light. I'm so impressed by the sights that I almost forget what I'm here for: to take photographs.

I want to get close to the sharks, so I venture away from the reef. But again, it's no use. The strong current whisks me up and over the reef crest, away from all the action.

I am alone now, not a good diving practice and not a particularly good

After dark, spiked sea cucumbers leave their lairs to feed. More than five hundred species of sea cucumbers have been documented. One species shelters the pearlfish, which enters its digestive track through the anus each day to avoid predators. When night falls, the pearlfish leaves its home to forage, returning the next morning to its host.

Facing *Sea fans grow at right angles to prevailing currents, placing their thousands of polyps in the best position for gathering plankton.*

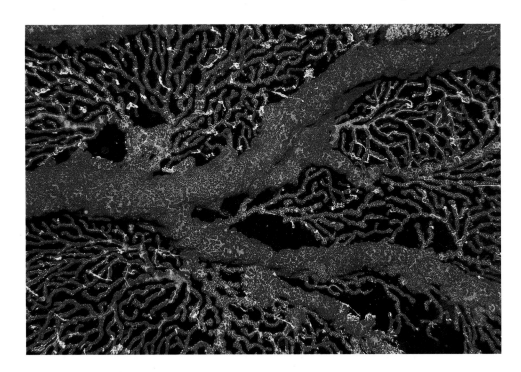

Sea fans and other soft corals are sometimes described as octocorals, because each polyp divides into eight sections: eight tentacles, eight stomach sections, and so on. Polyps in hard corals, or hexacorals, reveal an anatomy built on multiples of six.

Facing At Blue Corner, where the currents are swift and the water is exceptionally clear, schools of hundreds of fish are so dense they nearly block out the sun—a sign of a thriving marine ecosystem.

feeling. I have been at sixty feet for almost half an hour, so I can't surface immediately. I must make a safety stop at fifteen feet to decompress.

As I float, I keep one eye on my gauges and one on the lookout for sharks. The five-minute safety stop seems to last forever, but finally it's time to go up.

I surface in three-foot swells and driving rain. I do a 180-degree turn and finally see the boat—a half-mile away! There is no way the captain and crew can see me through the dense rain as I bob behind the waves. I inflate my six-foot orange diving safety sausage. Once it's inflated, I notice the boat turning toward me.

As I wait for the boat to arrive, my anxiety about floating in the water increases. That's where most shark attacks occur. I keep checking below, but there's not a shark in sight.

Back on board, I breathe a sigh of relief and prepare for the next dive. I change tanks, change film, and enjoy the fresh air for an hour. I can't wait to get back in the water for another exciting dive. The site: Blue Corner, of course.

Chandelier Cave, near Koror, gets its name from the huge limestone stalactites that hang from the ceiling. Inside the cave, dark-brown stalactites decorate two air pockets large enough to be explored on foot.

. . . .

GEOLOGICALLY, PALAU OFFERS MANY WONDERS, INCLUDING UNDERwater caves. Erosion and wave action first hollowed out these underground chambers in the porous limestone islands. Over time, as the sea level changed and tides rose and fell, stalactites formed.

Today, my dive guide Dexter and I are looking for Chandelier Cave, one of the most noteworthy underwater caves in Palau. Yesterday, at low tide, the entrance was easy to find in the dense mangroves that line the shore. Today, at high tide, the entrance is several feet underwater. After snorkeling for fifteen minutes, Dexter and I find it and begin our dive into darkness.

Cave diving is not for everyone, and especially not for the claustrophobic. We are slowly being swallowed up, the available light fading as we go deeper and deeper. I turn around every few seconds, making sure I can still see daylight.

I want to know where the exit is so I can find it in case of a problem.

We're only about thirty yards in when Dexter shines his light straight up. As I follow his beam, I see several stalactites hanging in the water. As I photograph these beautiful formations, I don't realize that I'm slowly ascending. Suddenly and unexpectedly, I break the surface of the water. I'm in an underground cave. Is it safe to breathe? I wonder. I take my regulator out of my mouth and take a small breath. The air is fresh.

Dexter breaks the surface. With our two dive lights, the small grotto is almost fully illuminated. Stalactites are hanging everywhere, some as much as six feet long. The walls have been beautifully shaped by wave action. I inflate my buoyancy compensator and float for several minutes. I'm enjoying the peace and serenity of the moment, but I'm also thinking about the thousands of tons of porous limestone above my head and our lone passage to open water.

We submerge and notice the faint light at the opening of the cave. It's nearly 5:00 P.M., and soon the sun will set, cutting off our guiding light. Slowly, we make our way back through the maze of chandeliers.

Big-eye squirrelfish use their large eyes to hunt small crabs, shrimps, and juvenile fish in dark nocturnal seas. During daylight hours, big-eyes rest under coral ledges and in caves, such as this vertical cave off Ngemelis Island. Sometimes they travel in schools, but at night, they are solitary hunters.

Left *Caves and tunnels riddle the limestone foundations of the islands of Palau.*

With its small population and strong traditional respect for nature, Palau has a good chance of preserving its extraordinary marine resources.

Facing Many species of starfish live in the warm waters of Palau. Although they have no eyes, starfish detect light and shadow with light-sensitive organs in their arms. This helps the animal find protected areas in the reef and perceive approaching predators.

ON THE LAST DAY OF OUR PALAU ADVENTURE, I'M IN A MELANCHOLY mood, as I often am after a fulfilling expedition. I'll miss the thrill of seeing something new and exciting on each and every dive, miss the Palauans who have become my friends, and miss giving conservation presentations to the local schoolchildren. I will not, however, miss seeing the local delicacy in the finer restaurants: fruit bat soup—served with the whole bat, floating face up, wings outstretched.

On this trip, our topside explorations were a fascinating short course in tropical flora and fauna. Our visits with local fishermen, government officials, and divers also gave us a glimpse into Palauan history and culture.

Best of all was the two-hour aerial tour of the islands that Susan and I made in a light plane. Cruising through the clouds, we could see just how untouched and undeveloped the outer islands are, unchanged over thousands of years. The tapestry formed by the islands and coral reefs is unlike any scene we have documented in our travels. I hope that thousands of years from now, Palau will still remain the natural wonder it is today.

I feel a bit uneasy sitting on the beach, watching the setting sun. This is our last Seven Wonders expedition; a year and a half of diving and exploration are behind us. We've been running at a frantic pace, planning expeditions, traveling to remote sites, sorting through thousands of Kodachrome slides, compiling data from scientists, and writing about our adventures.

For some time after the glowing red sun slips below the horizon and the sky fades from red to orange to yellow and then to dark blue, I wonder: what am I going to do now? What new adventures does the future hold? I think of our child to be, and realize that for Susan and me, this will be the greatest adventure of all. And we'll know we've succeeded on our Seven Wonders adventures if someday we can show our child some of the wonders we've sought to preserve.

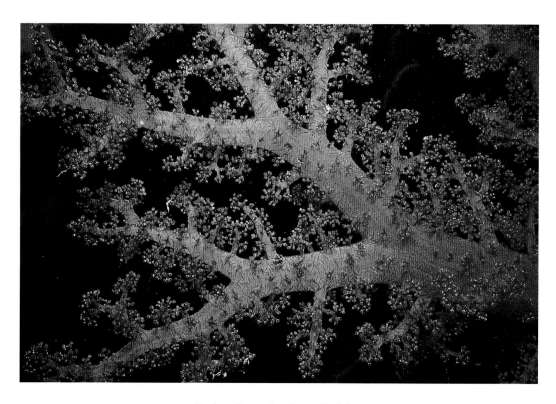

Pink soft coral colony, Red Sea

PHOTOGRAPHING UNDERWATER

· · · · · ·

You need five basic elements to get good underwater pictures: an interesting subject, dependable camera equipment, film that can record the true color of your subject, practice, and luck.

Interesting subjects are not lacking on a coral reef, where you'll find dozens of species of colorful fishes and corals. The trick is learning how to see underwater—how to identify potential subjects, many of which are masters of camouflage. Then you must learn how to compose and shoot before the scene changes, which can happen in the blink of an eye.

Also, read about animal behavior. This will give you a good idea of what to expect underwater in different parts of the world, at different times of day, during different tides, and at different times of the year.

A dependable camera with sharp optics is a must for the serious underwater photographer. Since 1979, I've been using the Nikonos 35mm system; currently I use the Nikonos V. I use the 15mm and 20mm lenses for wide-angle pictures. For macro shots, I use either the Nikonos close-up kit on the Nikonos 35mm and 28mm lenses or a Nikon N8008 with a 60mm macro lens in a Stromm housing.

Ninety percent of my underwater pictures are flash pictures. For these I use a Nikonos SB-103 flash (or two SB-103 flash units positioned off-camera for dual lighting). A flash brings out the true color and detail of underwater subjects. However, I also like to take pictures in natural light. These images may not have bright colors and sharp detail, but they do capture the reef as it appears to the unaided eye.

To keep my cameras in good condition, I rinse them thoroughly after each dive and soak them for a day in fresh water when I return home. Salt corrosion is one of the underwater photographer's worst enemies, and you must do everything possible to keep salt off your camera.

Topside, I use the Nikon N8008 and a selection of autofocus lenses. I almost always use a Cokin polarizing filter to reduce glare on the water and to darken white clouds against a blue sky.

Film is a personal choice, and five different underwater photographers may choose five different films. Personally, I like the warmth of Kodachrome 64 for most close-up and wide-angle pictures. I also find this film's fine grain structure holds up beautifully when enlarged, even when I have 6 x 9-foot LaserColor murals made.

For underwater pictures in natural light, I use Kodachrome 200, which provides a faster shutter speed to stop subject movement and a smaller f-stop for increased depth of field.

Topside, Ektachrome 100 is my standard film, and I use Ektachrome 50 when I have time to spend composing a landscape.

All my film is processed by Kodalux, a firm that specializes in slide film processing.

Practice makes perfect is an old cliché, but true nonetheless. When I started taking underwater pictures, I used to go to the local pool every weekend and practice my wide-angle shooting techniques. When I got home, I filled the bathtub with water and practiced my macro photography. These efforts paid off in the field, where you do not have the time to practice.

Luck is essential for getting eye-catching images. You simply are not going to get a picture of a whale shark, moray eel, or sea turtle if the animal does not happen to swim within shooting distance. You can improve your luck by diving often and by asking local fishermen and dive operators where the best dive sites are.

You can also improve your luck by working hard at your photography. Remember the old saying: the harder you work, the luckier you get.

A final note: be sure you always put the preservation of the reef ahead of your determination to get the best photograph. A careless photographer can do a lot of damage in very little time. Remember to take only pictures and leave only bubbles.

Conserving Coral Reefs

. . . .

As scuba diving becomes more popular, the underwater environment becomes increasingly threatened. When you are diving, please help to preserve the beauty and health of our underwater wonders—all of them.

Please do not sit, kneel, or stand on coral. It is fragile, breaks easily, and grows slowly. Never break off a piece as a sample or souvenir. Leave shells underwater, too; shell collecting has already badly depleted many mollusk populations.

Coral is very susceptible to infection when its outer skeleton is damaged. So when you are swimming over the reef, the basic rule is *look, don't touch*. And please watch where you are kicking.

Hol Chan Marine Reserve, Belize
***Facing** Reef crest on coral "bommie," Great Barrier Reef*

Being a competent, well-trained diver with good buoyancy control is one of the best ways to be sure you won't hurt the reef.

Read up on underwater ecosystems. Educate others to preserve the environment. Only with the help of all divers will our underwater world be preserved for generations to come.

For more information on how you can help protect the marine environment, you can write to the following conservation organizations. Most have newsletters that feature items of interest to dedicated conservationists. Some even have conservation programs and expeditions in which you can participate.

CEDAM International
One Fox Road
Croton-on-Hudson, NY 10520

Cousteau Society
8440 Santa Monica Blvd.
Los Angeles, CA 90069

Center for Marine Conservation
1725 DeSales Street NW
Washington, DC 20036

Earthwatch
680 Mount Auburn Street
Watertown, MA 02272

University Research Expeditions
University of California
Berkeley, CA 94720

Wildlife Conservation International
Bronx, NY 10460

Tunicates, Palau

A FEW WORDS OF THANKS

. . . .

Without the help and guidance of many talented individuals, this book would not have been possible. Here I'd like to give thanks to the conservationists and friends who have supported my efforts.

First, my thanks go to my best friend and diving buddy—my wife, Susan. For twelve years, she has been an invaluable partner, sharing highs like snorkeling with sea lions in the Galápagos Islands and lows like my extreme seasickness in the Philippines. Perhaps what I appreciate most about Susan is her never-ending positive attitude toward exploration and adventure, a quality that makes each expedition productive and memorable.

Other family members have also generously donated their time and talent. My father, Bob Sammon, was my constant adviser and has been since I was a kid. I distinctly remember him saying when I was about six years old, "You have two hands and a head like everyone else. You can do it." This perspective helped give me the confidence to undertake the Seven Wonders project.

My mother, Jo Sammon, read the drafts of each chapter and shared my enthusiasm for each adventure. Her Sunday phone calls encouraged me when I was swamped with the paperwork for this project, from visa applications and letters to government officials to travel itineraries and thank-you notes.

A warm "thank you" to the Russian scientists who kept our Lake Baikal ice diving hole open from dawn to dusk.

Joe Irwin, Susan's father, handled the project's many contracts, all in return for the satisfaction of helping someone with a good cause.

Paul Bush, known in Mexico as Don Pablo Bush Romero, is the founder of CEDAM International and CEDAM de Mexico. Since 1985, Paul has trusted me to guide the organization as its president and to lead CEDAM expeditions around the world, an experience which has brought me my most meaningful friendships and most exciting adventures.

I especially wish to thank the "Seven Underwater Wonders of the World" selection committee. The day in August 1989 when we chose the Seven Wonders was a high point in my life; I was enthralled by the presentations and awed to be in the same room with some of the nation's leading scientists, explorers, and naturalists.

The contributions of four committee members deserve an extra note of thanks. Emory Kristof of the National Geographic Society helped us mobilize our Lake Baikal expedition within months of the meeting—no small task. Lt. Jim Morris of the National Oceanic and Atmospheric Administration taught us how to use dry suits for the Baikal dives. Marsha Sitnik of the Smithsonian Institution worked a miracle and got us a rare copy of Mikhail Kozhov's *Lake Baikal and Its Life*, a book that proved invaluable in our quest for information on this remote, enchanting area. And Dr. Ernie Ernst of the New York Aquarium started it all in 1985 by sparking my interest in marine conservation.

Another person at the August 1989 meeting deserves credit for making this book happen. Frank Thomasson, president of Thomasson-Grant, liked my idea and saw the potential for an art-quality book on the subject. The *Seven Underwater Wonders of the World* is the result of his commitment and dedication to quality. Owen Andrews, my editor, polished the manuscript, which I wrote almost entirely on site before and after the dives. Leonard Phillips, the designer, brought photographs and words together with patience and artistry.

Along the way, several manufacturers provided products or services that helped us document the sites. For their support, I thank Catalina Cylinders, Nikon, Eastman Kodak, Kodalux Processing Service, LaserColor Laboratories, Sony, Stromm Underwater, and U.S. Divers.

I also extend my heartfelt thanks to individual supporters of the project: Carl Boyer, Mary Ann Delemel, Frank Fennell, John Hada, Jill Haines, Michael J. Kelly, Fran Leahan, Bruce Lisle, Dave Stancil, Michael D. Sullivan, and Glen Williams.

I owe an important debt to the authors whose work I've read over the past years as I've learned about the biology and ecology of all the world's underwater wonders. For readers who would like to expand their knowledge of marine life, a brief reading list follows these notes.

Two magazines have also provided many helpful insights: *National Geographic*, with its frequent and superb articles on underwater life, and *Oceanus*, published by the Woods Hole Oceanographic Institution. *Oceanus'* special issues on the Caribbean, Australia, the deep ocean vents, and the Galápagos Archipelago are particularly worth noting.

My final thank-you goes to the individual who first showed me the wonders of the underwater world, Jacques Cousteau—a true pioneer in the field of marine exploration.

R.S.

RECOMMENDED READING

Herbert R. Axelrod and Cliff Emmens, *Exotic Marine Fishes* (Jersey City, N.J.: T.F.H. Publications, 1988).

Jacques Cousteau, *The Ocean World* (New York: Harry N. Abrams, 1979).

Richard Chesher and Douglas Faulkner, *Living Corals* (New York: Clarkson N. Potter, 1979).

Paul Human and Ned Deloach, *Reef Fish Identification* (New World Publications, 1989).

Victoria A. Kaharl, *Water Baby: The Story of Alvin* (New York: Oxford University Press, 1990).

Eugene H. Kaplan, *A Field Guide to Coral Reefs of the Caribbean and Florida* (Boston: Houghton Mifflin, 1982).

Franz O. Meyer, *Diving and Snorkeling Guide to Belize* (Houston: Pisces Books, 1990).

Mikhail Kozhov, *Lake Baikal and Its Life* (Dr. W. Junk Publishers).

Joseph S. Levine and Jeffrey Rotman, *Undersea Life* (New York: Stewart, Tabori & Chang, 1986).

Les Line and George Reiger, *The Audubon Society Book of Marine Wildlife* (New York: Harry N. Abrams, 1980).

Godfrey Merlen, *A Field Guide to Fishes of the Galápagos Islands* (Wilmot Books, 1988).

David Pilosof and Lev Fishelson, *Mysteries of the Red Sea* (Masada Publishers, Ltd., 1983).

John E. Randall, *Caribbean Reef Fishes*, 2nd ed., rev. (Neptune City, N.J.: T.F.H. Publications, 1983).

C. Richard Robins and G. Carlton Ray, *A Field Guide to Atlantic Coast Fishes* (Boston: Houghton Mifflin, 1986).

Early morning in the Red Sea

INDEX

Arrow crab, Belize